TEACHING
HANDBOOKS

Teaching Primary Mathematics

John Suffolk

MACMILLAN

Macmillan Education
Between Towns Road, Oxford OX4 3PP
A division of Macmillan Publishers Limited
Companies and representatives throughout the world

www.macmillan-africa.com

ISBN 978-0-333-75016-2

First published 2004

Typeset by Jim Weaver Design Ltd
Illustrated by Tek Art
Cover design by Charles Design Associates
Cover photographs by Sally Luck

Dedication

This book is dedicated to my wife, Nellie, who began as a Primary school teacher in Zambia and is now teaching English at the University of Brunei. She read the first draft, and has supported and tolerated me as I developed it. Without her beside me, I would do very little, and without her I would not have the wonderful children Stephen, Angela, Shupiwe and Christine.

Printed and bound in Malaysia

2008 2007
10 9 8 7 6 5 4 3

Contents

Prefaces

It has been observed that, for many people, their early encounters with mathematics induce fear and resentment. Their memories of mathematics often involve remembering long chains of unrelated figures and formulae and other apparently meaningless facts. At the same time, teachers in primary schools face many challenges. Primary school teachers in Africa have to handle large classes. They have to search for teaching materials, which often are not readily available. If teachers closely follow the advice and methods given in this book, we shall soon find mathematics becomes one of the more favoured and enjoyable subjects in the primary school curriculum.

When mathematics teachers construct their own lessons, it helps them to understand each topic fully before they present it to the pupils. The teacher can also choose teaching methods that can be shown the pupils practically, and set in their own environment. This is the only way in which children will be able to see mathematics as meaningful and relevant to them. *Teaching Primary Mathematics* is written as a resource book for all teachers of primary mathematics. The author urges teachers to develop their own teaching materials. This can help resolve a number of basic needs for teachers. In the first place, teachers will find material that they construct themselves to be meaningful and appropriate to the pupils they teach. Secondly, as teachers contribute resources to their schools, the lack of materials in schools will ease. Thirdly, pupils using appropriate practical approaches can discover more mathematics than those just reading about it. This will enable pupils to become active participants in the mathematics they are learning, which will give them a solid foundation for the future.

Teaching Primary Mathematics is divided into two main parts. Section A deals with methodologies of teaching mathematics. The second part (Sections B to E) is about mathematics topics. This will help trainers to handle both the methods and content effectively in helping teachers to become constructors of their own teaching strategies. This book will definitely be a useful tool to many people who are concerned with the teaching and learning of primary mathematics. It will prove an invaluable resource in primary teachers' colleges, both pre-service and in-service. Serving primary school teachers can use this book to improve and update their teaching methods. Mathematics educators, inspectors, and advisors will also find *Teaching Primary Mathematics* a good reference book.

Justin Kapya Shipola, Head of Mathematics
National In-Service Teachers' College, Chalimbana, Zambia

There is agreement worldwide that mathematics is not only important but it is very useful, and for these reasons every child should study mathematics at school. Child-centred methods are ignored to the disadvantage of the growing child. Sadly, the fact is that many pupils do not enjoy doing mathematics.

Teaching Primary Mathematics will help to overcome this. It provides insights into the processes of teaching and learning mathematics by focussing on a wide range of issues and topics. The author has included Mathematics Methods and all the Basic Topics that can be found in any primary mathematics syllabus. It is a resource book that aims at developing mathematical thinking by providing a series of practical ideas in mathematics lessons. The new thinking in this book is that as the teacher allows the pupils to investigate, to explore, to discuss and to relate mathematics to real life situations. Mathematics becomes functional in their lives.

In recommending this handbook to every teacher and to every teacher trainee I wish to assure you that it is a book that will provide you the opportunity to develop the knowledge base, attitudes, and skills of your pupils, and enable you to reach out for the resources, both mental and physical, to make their mathematics activity fulfilling.

Daniel O Apronti, Mathematics Educator
Accra Training College, Accra, Ghana

Introduction

Macmillan Teaching Handbooks

This series has been designed for student teachers and practising primary teachers in African schools. It has been developed as a result of consultation with practising teachers, student teachers and college teacher trainers. The series includes one book covering all that the aspiring or busy teacher needs to know about the basic methods of primary teaching, and one book for each of the main subjects: English, Science, Mathematics, and Environmental and Social Studies. All the books cover upper and lower primary.

The books are designed to be used by unqualified or newly qualified teachers, by students studying on formal college courses and by experienced teachers who wish to improve their practice. Readers are encouraged to try out methods and strategies to improve learning and to reflect on their own teaching, either individually or in groups. The material can also be used in Teacher Training College courses or equally effectively in school-based staff development programmes supported by the headteacher, teachers themselves or by primary school advisors.

The books present the material in an attractive, easy to use form, without compromising on concept level, coverage or depth. They make no assumptions about previous study or knowledge, avoid jargon and use clear language. They provide a wealth of practical suggestions and ideas, keeping in mind the problems and challenges that teachers face in the classroom.

Teaching Primary Mathematics

Maths is often taught badly. Many children grow with memories of hating maths at school. They remember boring lessons in which they learnt formulae and tricks and had to answer endless problems to which they saw little point. Maths does not have to be like that. It can be made much more interesting, more easily understood and, even, fun.

The purpose of this book is to help teachers to teach primary school maths in a way that is understandable and enjoyable, both for the pupils and the teachers. The book is set out so that a busy teacher can easily find the help he or she needs. Students who are preparing to be teachers of maths can also use the book as part of a pre-service or in-service programme of study.

Layout

The book is divided into five sections:

A **Mathematics methods**
B **Mathematics topics — number and algebra**
C **Mathematics topics — handling data**
D **Mathematics topics — measures**
E **Mathematics topics — shape and space**

Section A looks at general issues of teaching primary maths. It covers subjects such as how to teach, planning teaching, questioning in the maths classroom, and so on.

Sections B to **E** look at the teaching of the various topics in the primary school syllabus. These sections are set out so that a teacher wanting to find quickly some ideas on how to teach (for example) the addition of fractions can easily locate that topic in the Contents. Busy teachers do not have time to read an extensive discussion of issues relating to how to teach a particular topic. They want to be able to find teaching ideas easily and quickly. This book is designed specifically to enable them to do that.

Within each chapter there are some suggested **Activities**. These can be done either by a teacher or by a group of teachers in a school or in a workshop, or by students in pre-service training.

Acknowledgements

A book such as this depends on many ideas accumulated over the years and whose source is often forgotten. The *Maths for the Majority* series of some thirty years ago was, for me, an early source of excellent ideas. Many ideas have come from the presentations and assignments of students I have taught over the years, especially those at the University of Brunei. I have learnt a lot from working with many fine colleagues. In particular, I would like to mention:
the late Kingsley Manzi Wachila, a former teaching colleague in Petauke, Zambia;
the late Roger Edwards, formerly a colleague at the University of Papua New Guinea;
David Womack and Peter Pool, former colleagues on the Instant Project in Namibia;
Martyn Quigley, formerly of the University of Brunei Darussalam, now at the National Institute of Education in Singapore;
Hjh Zaitun bin Haji Taha of the University of Brunei Darussalam;
and especially Ken Clements, professor of Mathematics Education at the University of Brunei Darussalam.

I also acknowledge the patience of staff at Macmillan Publishers, notably Tom Hardy, as they waited for a very long time for this book to materialise.

Bibliography

There are many more good ideas about the teaching of maths than can possibly be explained in a book such as this. Teachers of maths are encouraged to read as many books and magazines as possible to learn new ideas and improve their teaching. Books that are particularly useful for primary school teachers include:

Askew, M., *Teaching Primary Mathematics* published by Hodder and Stoughton, Abingdon, UK in 1998. There are a lot of good ideas and reports of research into primary school maths.

Atkinson, S. (ed.), *Mathematics with Reasons* published by Hodder and Stoughton, Abingdon, UK in 1992.

Lovitt, C. and Clarke D., *The Mathematics Curriculum and Teaching Program Activity, Volumes 1 and 2* published by Curriculum Corporation, Canberra, Australia in 1988. These books contain a marvellous collection of ideas for teaching maths.

Thompson, I. (ed.), *Issues in Teaching Numeracy in Primary Schools*, published by the Open University, Buckingham, UK in 1999.

Williams, E. and Shuard, H., *Primary Mathematics Today* published by Longman, Harlow, UK. A large book with many helpful suggestions.

Magazines such as *Mathematics in School* and *Mathematics Teaching* from the UK, *The Arithmetic Teacher* from the USA and *Vinculum* from Australia. All contain excellent teaching strategies.

Section A Mathematics methods

A1 Teaching with justification

1 Introduction

This unit sets the tone for the rest of the book. The teaching of maths is discussed in the belief that maths should not be taught as a body of facts that pupils have to remember without any understanding. Maths taught as an incomprehensible set of rules and methods is boring and not very successful — look at the school results and the number of people who are proud to say that they are no good at maths.

Maths exams do not ask pupils to remember a lot of things. They ask pupils to solve problems. Learning maths well mostly requires intelligent learning. Skemp borrowed the suggestion of Mellin-Olsen that there are two types of 'understanding.'

'Relational understanding' is *knowing both what to do and why*.
'Instrumental understanding' is *knowing rules without reasons*.

He gave some examples:

- area of rectangle $A = l \times w$. Many know the formula but do not know where it is from.
- 'turn it upside down and multiply' for division by a fraction. Why?
- the circumference of a circle = πd. Why?

2 Justifying mathematics

Nothing is true in maths simply because a mathematics teacher says it is true. It is possible to **justify** (show that something is reasonable) everything taught in mathematics. Every result, property and method can be justified in some way. This is true for all areas of mathematics.

The emphasis is on **showing**. The pupils have to be shown by the teacher that what they are being taught makes sense. Pupils should understand why what they are being taught is true. This means that the methods used have to be appropriate for the age of the children being taught. The abstract methods of higher mathematics are not appropriate for Primary school children. Young children need practical methods.

3 Some examples

Here are some examples of practical approaches to helping pupils understand what they are learning relationally.

Area of a rectangle

It is well known that the area of a rectangle is found by working out **length** × **breadth**. Many of us were taught this as a rule and are unsure as to how it might be justified, yet it is a rule that even Primary school children can discover for themselves.

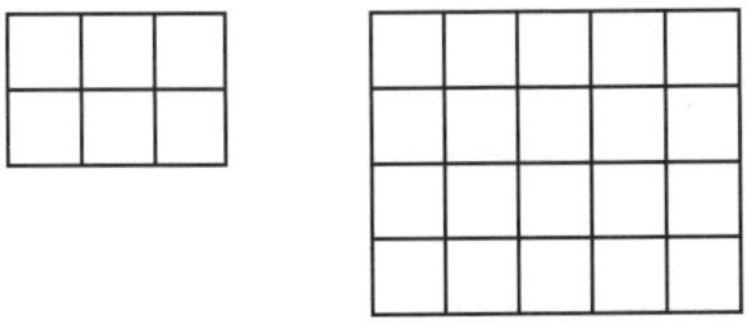

Look at these rectangles:
How many squares are there in each of them?
We count the squares.
The 2 by 3 rectangle contains 6 squares.
The 4 by 5 rectangle contains 20 squares.
Is there a connection between the number of squares and the length and breadth of the rectangle? The pupils should be able to find out the rule for themselves and tell the teacher.

The circumference of a circle

Pupils learn that $C = \pi d$ and that π is *about* 3.14 or $\frac{22}{7}$, but most appear to have no idea where this formula comes from. Yet it is quite a simple matter to demonstrate with a cylindrical tin and a piece of string that C is a little more than 3 times the diameter. This method is appropriate justification for Primary school children. A more accurate calculation of π is only possible much later on.

Sum of the angles of a triangle

Pupils can discover the sum by drawing a triangle and measuring its angles. Add the angles up — and what do the pupils notice? The angle sum is 180°. This is true for every triangle they draw, and discovering this rule is only limited by their ability to use a protractor accurately.

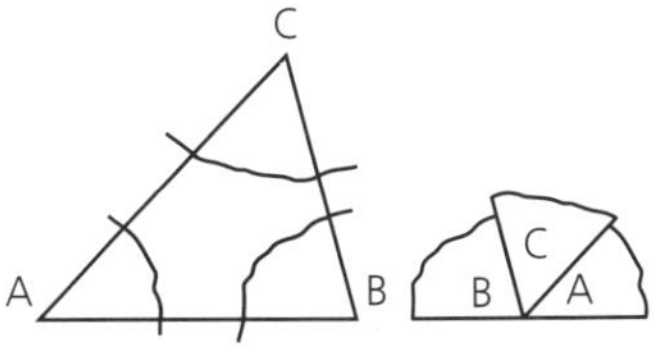

Another convincing method is to draw a triangle and tear off the corners. Put the three angles together — and what do the pupils notice?

The angles form a straight line — 180°.

4 The rules for dividing fractions

Everybody learns that to divide by a fraction you 'turn it upside down and multiply'. For example,

$$\frac{5}{8} \div \frac{3}{4} = \frac{5}{8} \times \frac{4}{3}.$$

Why is that? What is the reason for it? Most people learn the rule at school without understanding why.

There is a quite simple explanation. Let us rewrite the expression as

$$\frac{5}{8} \div \frac{3}{4} = \frac{\frac{5}{8}}{\frac{3}{4}}.$$

This looks very messy, partly because of the fraction in the denominator. How can the denominator be simplified? First, remember that if we multiply the denominator by its inverse, it becomes 1. Also, if we multiply the numerator and the denominator by the same number, the value of the fraction does not change. So multiply the numerator and denominator by $\frac{4}{3}$ (the inverse of the denominator).

Now the expression will look like this:

$$\frac{5}{8} \div \frac{3}{4} = \frac{\frac{5}{8}}{\frac{3}{4}} = \frac{\frac{5}{8} \times \frac{4}{3}}{\frac{3}{4} \times \frac{4}{3}}.$$

This looks quite frightening but we can work with it a little more to give:

$$\frac{\frac{5}{8} \times \frac{4}{3}}{\frac{3}{4} \times \frac{4}{3}} = \frac{\frac{5}{8} \times \frac{4}{3}}{\frac{\not{3}^{1}}{\not{4}_{1}} \times \frac{\not{4}^{1}}{\not{3}_{1}}} = \frac{\frac{5}{8} \times \frac{4}{3}}{1}.$$

Finally, remember that any number divided by 1 is also unchanged, so the expression simplifies to:

$$\frac{\frac{5}{8} \times \frac{4}{3}}{1} = \frac{5}{8} \times \frac{4}{3}.$$

This is fairly sophisticated mathematical reasoning, but reasoning that older Primary children are capable of understanding. The calculation only has to be done once in this way to explain the 'rule'.

There is another method discussed in Section B, Topic 15.

Activity

1 The geometrical justifications on page 3 use practical experiments. Find other experiments used in this book to demonstrate the truth of geometrical properties.

2 Justify the following procedures in arithmetic:
- 'borrowing' in subtraction
- multiplication of fractions.

A2 Teaching mathematics constructively

1 Introduction

Jean Piaget believed that children, in the words of Papert, 'are not empty vessels to be filled with knowledge ... but active builders of knowledge.' Piaget wrote that 'Children have real understanding only of that which they invent themselves, and each time that we try to teach them something too quickly, we keep them from reinventing it themselves.' Inspired by Piaget's ideas, it has become popular to argue that maths teachers should help their pupils to build or **construct** their own understanding of maths. As Vygotsky has stressed, the construction of understanding takes place in a social and cultural context that affects how that understanding develops.

2 Teaching constructively

Children do not receive knowledge passively. They have to build knowledge up actively. The world is learnt about through experience. Constructivist teachers want pupils to build up understanding for themselves. This is in contrast with traditional, transmission views of learning as a process by which the teacher gives pupils skills.

The constructivist teacher is interested in the thinking that is going on inside the pupils' heads and helping the pupils make sense of the world. The **errors** that pupils make are very interesting as they reveal what help the pupils need.

For this sort of teaching, it is very important that there is good social interaction in the classroom. There needs to be a good relationship between teacher and pupils.

An example of teaching constructively

Suppose a teacher wants to teach her class the formula for the area of a right-angled triangle. The pupils already know the meaning of the words 'triangle', 'height' and 'base'. The following illustrates how the lesson might go.

1 The teacher decides on the approach that she is going to use. She wants the pupils to think of a triangle as half a rectangle and construct the formula from that. So she decides to give the pupils paper rectangles (she could have asked the pupils to cut them out, but this saves time in the lesson), and she needs some scissors for the pupils to cut them up.
  She starts the lesson by asking some questions around the class to check that the pupils know:
  - what a triangle is — three sides, and so on
  - what are the base and height of a right-angled triangle
  - what area is, how to find the area of a rectangle
  - multiplication number facts.

2 The teacher gives each pupil a small paper rectangle. '*What can you tell me about it?*' The pupils give various answers — its name, it is the shape of a face of a brick, its area, its angle sum, and so on. These answers are praised and written on the board.

3 She tells them to draw a line along the diagonal and asks them, '*Can you fold your rectangle into two triangles?*' They will tell her to fold along a diagonal (although they may not know that word).

4 The teacher gives the pupils the scissors and asks them to cut the rectangle in two. She asks, '*What can you tell me about the triangles?*' They answer that the triangles are right-angled and that they are the same. '*What do you mean by the same?*' They may say that the triangles have the same angles, and so on. Hopefully they will say that their areas are the same.

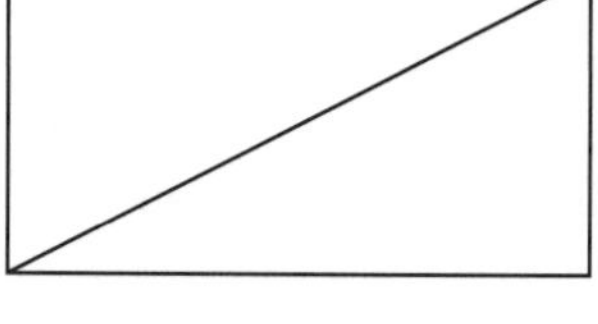

5 She asks, '*Which is more, the area of the rectangle or the area of one of the triangles?*' They should reply, 'the rectangle'. '*How much larger?*' Twice as large.
She draws a rectangle on the board with the diagonal drawn.
She writes:

area of rectangle $=$ length $\times$ width $=$ 2 $\times$ area of triangle
area of triangle $=$ ... area of rectangle
[The pupils tell the teacher to put $\frac{1}{2}$ in the gap.]
$= \frac{1}{2}$ length $\times$
width
[The teacher explains how the length and width of the rectangle could be seen as the base and height of the triangle.]
$= \frac{1}{2}$ base $\times$ height

Activity

1 Design a series of questions and activities that you could use to enable pupils to develop an understanding **for themselves** of the following topics:
- the relationship between the circumference and diameter of a circle
- adding two fractions (sum less than one)
- multiplying two negative numbers together
- finding the angle sum of a polygon.

2 **Role-play** Turn the questions into a play with some of you pretending to be pupils and some being the teacher trying to teach constructively.

6 The pupils have told the teacher the formula. She just asked the correct questions. Now the pupils can apply the formula to find the areas of right-angled triangles, before discussing other triangles.

The conversation above is an idealised one, in the sense that the pupils always give the required answers. This takes practice and may not always work. The teacher should encourage the pupils, even if their answers are wrong, and even if they need more help than is needed in this story.

A3 Planning a mathematics lesson

1 Introduction

Every maths lesson should be properly prepared. Teaching is so much better when it has been well planned. This unit looks at the planning process.

2 Factors to consider when planning a lesson

Decide what is going to be taught and how it is to be taught. Several factors need to be thought about.

(a) Pupils' previous knowledge What must the pupils know *before* they learn the topic? Learning maths often depends on an understanding of what was learnt before. This is one of the main causes of the difficulties pupils have with maths.

Example 1
In order to find the angle sum of a triangle by measurement, pupils need to be able to:

- draw a triangle accurately
- add two- and three-digit numbers
- measure angles accurately with a protractor.

Example 2
In order to work out 71 – 43, pupils need to be able to:

- label columns T and O (or U)
- use subtraction facts
- take bottom ones digit away from top ones digit
- if it 'won't go', decompose a ten from the tens column in the top number
- take bottom ones digit away from new top ones number (subtraction facts)
- take bottom tens digit away from new top tens digit
- check answer is close to their estimate.

Activity

Make a list of what pupils need to already have learnt before they learn:

- long division problems such as 405 ÷ 27
- the formula for the circumference of a circle.

(b) Materials What materials are needed and how much? Are the materials readily available? Do you need to plan getting the materials well in advance? For example, in order to teach $C = \pi d$, a class of 30 pupils working in pairs would require 16 tins, 16 rulers and 16 pieces of string (one for each pair plus one for the teacher). If you are teaching about timetables, you can go to a travel agent or airline office and pick up some timetables several days before you teach the topic.

These materials do not need to be very expensive. Many of the most effective teaching aids can be made from left-over items such as matchboxes, empty cartons and other containers, and used paper. See Section A, Unit A14 for a discussion of this.

(c) Class ability and experience The range of ability and experience in the class will determine the depth, speed and approach used. Ask yourself what approach will work best with the class. Is the class mixed ability? Will less-able pupils need different activities?
Teaching mixed ability classes is one of the difficulties of teaching maths. See Section A, Units A6, A7 and A12 for discussions of this problem.

(d) Difficulties teaching and learning the topic Ask yourself, *'Do I understand the topic fully?'* A teacher must know the topic well. There may be some tricky points. Ask yourself, *'Can I do all the questions in the textbook? Is there anything that I do not fully understand which I should ask my colleagues about?'*

Activity

Decide which topics in the syllabus you find difficult. Think about them and try to solve your problems by:

- discussing them with colleagues
- looking in books for explanations
- asking other people to help you.

3 Deciding how to teach a topic

(a) What is the best approach for your class? Before you decide how you will teach a particular topic, ask yourself some questions. *What is the approach used in the textbook? Is it a good way? What is the approach used in this book? Is it the same as the one in the textbook? Is it better? Is there a better way of teaching the topic? Will my pupils understand this way?* Look in books and articles and discuss with colleagues the best way to teach a topic.
Think about the important questions you should ask the pupils. These are called **key questions**. Make a note of the questions.

Activity

1 Think about a topic you are going to teach:
- what approach will you use?
- decide what are the **key questions** that should be asked in the lesson
- prepare the teaching aids needed.

2 **Role-play** Present the lesson to your colleagues as they pretend to be the pupils.

This activity can be undertaken for any topic in Sections B to E of this book.

(b) Is there anything that will be difficult to explain easily? The teacher should always go into the classroom confident that he or she can teach effectively. That is why it is so important to know that you understand the topic that you are going to teach. Discuss with colleagues what difficulties they have found.

(c) Think about the difficulties that the pupils may have. Are there any difficulties the pupils may have with the topic? Think about **your** pupils and plan how to help them overcome any difficulties. Remember your own school days and the difficulties you had with this topic. Find out from colleagues the difficulties that their pupils have had with the topic. Read books and articles for ideas. Look at old examination papers to see the standard that the pupils will be expected to reach.

Think about the level of English language that your pupils have when you prepare your lesson. Make it as simple as possible. (See Section A, Units A9 and A10.)

Activity

Choose a topic and consider the difficulties that pupils have when learning it. Try to find out the sort of mistakes that pupils commonly make, from books and articles, and from other teachers. Consider what can be done to prevent and correct those mistakes.

This activity can be undertaken for any topic in Sections B to E of this book.

A4 Writing a lesson plan

1 Introduction

A teacher should write down the preparation in a plan for *each* lesson. Lesson planning is something done throughout one's teaching life because a plan improves teaching. Here are some general remarks about maths lesson plans:

(a) The lesson plan is a guide, not a boss The teacher does not have to teach everything on the plan. There may be too much in the plan. It is better not to complete the work on the plan than to rush through it too quickly. Teachers sometimes have to forget about the plan if old work has to be revised extensively.

(b) When planning, consider what the teacher and children will be doing What approach is to be used? Is the method in the textbook good, or is there a better one? Is the method one that the pupils can easily understand? Do they spend too long doing the same thing? Pupils should have a variety of activities.
The teacher should not talk too much. The more active pupils are, the more interested they will be, and the more they will learn. (This is especially true of boys.)

(c) Think very carefully about the numbers that are to be used in problems Changing a number in a problem very slightly can change the problem significantly. Primary pupils can calculate 34 – 21, but not 21 – 34. Adding $\frac{1}{5} + \frac{2}{5}$ is easier than adding $\frac{1}{5} + \frac{2}{7}$.

(d) Choose examples carefully Consider the examples that the pupils will do. Often, teachers do questions 1–3 from the textbook on the board; pupils do questions 4–8 in class and questions 9–12 are done at home. This means the teacher does the easiest questions, pupils do the middle ones with the teacher, and then the hardest ones at home on their own! Mix them up — some easy and some difficult questions orally, and some easy ones (to encourage all pupils to have some success) and some harder ones (to challenge brighter pupils) as homework.

Should the teacher use the examples in the textbook? Usually it is better to make up your own examples so that the pupils can look at the teacher's examples **and** the examples in the book. Then they have more examples to help them understand.

2 Suggested lesson plan format

Teachers often start a maths lesson by reviewing previous work. Then they introduce a new concept or skill, do some worked examples on the board, and then set the class some exercises. This pattern will not always be followed (in fact it should not be followed **always** because there should be variety in teaching).

Whatever the lesson, the plan should have the following parts:

1 Headings
2 Lesson objectives
3 Pupils' difficulties
4 Materials and equipment needed
5 Introduction (state the time needed)
6 Development (state the time needed)
7 Conclusion (state the time needed)
8 Pupil evaluation
9 Evaluation of the lesson

Consider the amount of **time** needed for parts 5, 6 and 7. Make sure that any activity you plan can be done in the time available.
Let us look at each of these parts in more detail.

1 Headings Your name, class, school, time and date of the lesson, topic, etc.
Why is this on the plan? For record purposes, to help anyone see at a glance the details of the pupils, the teacher and the topic.

2 Lesson objectives What you hope to achieve.
Why is this on the plan? Every lesson should have a purpose. Think about the purpose. The objective of a lesson may be that the pupils will be able to do something at the end of the lesson that they could not do before. But that is not all of it. Many lessons will encourage the pupils to think for themselves, to express themselves, to improve their problem solving ability, and so on. It will also be an objective that they enjoy their learning.

3 Difficulties pupils may have You will have researched what common difficulties arise.
Why is this on the plan? What are the difficulties the pupils may have with this topic? If these are considered in advance, it may be possible to teach so that the pupils avoid them.

4 Materials and equipment needed Some lessons require more than books and chalkboards.
Why is this on the plan? To help the teacher think about the materials to be taken to class which have to be prepared in advance. The plan gives you a checklist of these things. (Do not list items like 'chalk' and 'red pen' that are taken to every lesson.)

5 Introduction (time to be stated) How you will start the lesson.
Why is this on the plan? Decide how to introduce the lesson. Often, there are many ways — a game, a review of previous work. Try to make it interesting. It could be a challenge, such as '*How could we find out what the angles of triangles add up to?*' It could be a **revision/extension** exercise, **revising** the work of an earlier lesson and then **extending** into the topic of the new lesson. List on your plan the key questions that you are going to ask.

6 Development (time to be stated) This contains the main points of the lesson.

Why is this on the plan? This is the heart of the lesson. It is often the first part of the plan that is prepared. It determines the materials and equipment needed and it determines the introduction. Think very carefully about the **approach** to be used — the textbook approach or another way? Again, list on your plan the key questions that you are going to ask in the lesson.
Consider carefully the **examples** to be used. Include in the plan all the examples to be used for:

- oral work by the class and teacher
- class work by pupils
- homework.

7 Conclusion (time to be stated) How you will end the lesson.
Why is this on the plan? This should sum up the lesson. It should briefly cover the ground covered in the lesson. Make sure most pupils understood enough to be able to do the problems set for homework. Leave time for the pupils to make a note of homework. It is also the time to collect equipment that has been used, and tidy the room. Do not leave it until the last minute.

8 Pupil evaluation Do the pupils understand what has been done?
Why is this on the plan? The teacher needs constantly to evaluate his or her pupils, either formally through homework and tests, or informally through questioning and observation. This part includes the items you will use to evaluate the pupils. It will consist of questions from the textbook (chosen in advance) or from elsewhere.

9 Evaluation of lesson Did the lesson work?
Why is this on the plan? This is done after the lesson and should be an honest appraisal of the lesson. It is a chance to learn, reflect and improve. Teachers are very busy, but it is important to find time for this. Think about how the lesson went, and why. Jot down notes to remind you for next time you, or another teacher, teaches that lesson.

Activity

Write a lesson plan for the teaching of any topic that you choose.

Discuss the plan with your colleagues.

A5 Questioning in the mathematics classroom

1 Introduction

Both the teacher and the pupils should enjoy the environment in the maths lesson. The pupils must be involved in what is going on. The teacher must interact with the pupils. Questioning is one of the teacher's most powerful tools.

2 Why ask questions?

The purposes of questioning in the maths lesson are:

- to help pupils to learn
- to enable them to overcome difficulties
- to review previous work and evaluate progress
- to control and organise the class.

(a) Learning To help pupils learn, questions should be asked with individuals in mind. Questions like '*Do you understand?*' addressed to the whole class serve little purpose unless you have in mind a particular pupil who will answer the question. The way to ascertain if pupils have understood is to ask specific questions of individuals.

When the teacher does this, he or she has to be careful not to threaten the pupil. Try to avoid asking weaker pupils difficult questions — give them the kind of question that you think they should be able to answer. They will then be able to participate in the classroom discussion in a successful way, and that can only serve to improve their attitude to mathematics. For example, when working out 24×23, you might ask the weaker pupils what 3×4 is, and stronger pupils how to set out the calculation.

In any lesson, there are certain key questions to ask. These are the main extension questions that the teacher should ask. They are the focus of the lesson.

Key questions when teaching $C = 2\pi r$

Already revised

Meanings of words like 'circle' and 'radius'.

Extension questions

Which is longest: circumference *C*, radius *r*, or diameter *d*?

What is the relationship between *r* and *d*?, *d* and *C*?

How can we find out?

What should we measure?

Activity

What are the key questions to use when teaching these topics?

(a) multiplication of a two-digit number by one-digit number
(b) angle properties of isosceles triangles
(c) adding two mixed numbers

(b) Learning difficulties Questions can be used to help pupils avoid the learning problems that you know may arise. For example, you may be adding 54 + 238 on the board. You know that some pupils will set this out wrongly so you can anticipate this and ask them about the importance of setting out questions, as you teach.

(c) Evaluating progress Asking questions around the class as you teach a new topic will tell you whether the pupils have understood. Questions will bring out misconceptions and you will be able to address them as soon as they occur. Questions can be used to evaluate performance and to see whether all the individuals in the class have understood. Answers given as a chorus tell you very little.

(d) Controlling discipline If a pupil is not paying attention, or is misbehaving, then a question directed to him/her can control his/her behaviour. A question such as '*John*,' (pause), '*What is the angle sum of this triangle?*' will get his attention. (A simple question is more effective as the pupil has been brought back into the class activity but has not been humiliated or made to lose face.) Usually questions should not be preceded by a name so that all pupils think that they might be expected to answer.

Questions are also used to organise the class. 'Have you all brought rulers?' 'Are you ready?', and so on.

3 Types of questions

Most questions require the pupils to recall something:

What is a quadrilateral?'	(Answer is a definition)
What do you call a whole number divisible by two?	(Answer is a name)
Is the angle sum of a triangle 360°?	(Answer is a simple Yes or No)

Some questions (or requests) require **convergent** thinking, such as: *Solve* $2x + 3 = 5$.

Convergent thinking questions are at a higher level than memory questions. The pupil has to be able to use given or remembered information. There may be one expected answer, often found by carrying out a known procedure. (Sometimes several answers are possible, or the answer may be a description rather than numerical.)

Other questions are procedural, and about lesson organisation.

Some questions require thinking. They are **open** questions, to promote discussion and **divergent** thinking. For example, '*Should calculators be allowed in Primary school?*' Divergent questions ask the pupil to be original and think for him/herself, to organise and synthesise ideas, to infer and to predict. For example, '*You have all drawn triangles and measured their angles. What can you conclude?*', '*Add up several pairs of odd numbers. What do you notice? Can you prove this?*'

4 Some general points about questioning

1 Try to make your questions unthreatening. Pupils may be feeling unsure. Do not make them feel unsafe and inadequate with inappropriate questions.
2 Try to be skilful when you reject a wrong answer to avoid embarrassment. Encourage rethinking and improved answers. Praise all good answers. Try to find something good in wrong answers.
3 Use simple vocabulary, short sentences, slow and clear speech.
4 Ask questions that require *more than one-word answers.* Teachers who expect pupils to express ideas develop their thinking skills and their language. Frequently, too many questions in maths lessons ask for one-word answers. Try to ask questions that require answers that are more complex.

A useful procedure for asking questions is:

- ask the question of the whole class (but know which pupils you are directing the question to)
- *pause* after asking a question for 3–5 seconds to allow time for all the pupils to think about the question and formulate a response. Increasing the length of the pause has been found to:
  - increase responses
  - increase weaker pupils' participation
  - increase the number of questions asked **by** the pupils
  - decrease failures.
- *name* the pupil that you want to answer the question. If the question is an easy one, choose one of the less-able pupils. If it is a difficult question, choose one of the more-able pupils.

> ***Activity***
>
> Prepare some questions that require divergent, thoughtful responses that could be asked in Primary school. Make sure that the language is appropriate.

Some pupils will have their hands up and want to be chosen all the time. Some will have their hands down. Do not always choose the pupils with their hands up. If you do, lazy pupils will be able to have a quiet life by never putting their hand up. On the other hand, if you always ask for the answer from those who have not put their hands up, then lazy pupils will always raise their hands in the expectation that that ensures they are left alone!

A6 Varying approaches to mathematics teaching

1 Introduction

Too often maths is taught in a boring teacher-centred way, with the teacher describing how to solve a type of maths problem, doing a few examples on the board, and then setting an exercise for the pupils to do. This approach is then repeated lesson after lesson and is not very interesting. It also discourages many pupils. Maths can be taught in many ways. Do not teach each maths lesson in the same way.

2 Reasons for a variety of approaches

1 Teaching of maths should not follow a uniform pattern. There are several reasons for this. Teachers have an obligation to try and make the maths they teach interesting and understandable, especially to less-able pupils. Teaching all lessons in the same way is not very motivating. There needs to be variety to make lessons more interesting.

2 Different pupils have different abilities, backgrounds and interests. Teaching in one way may suit some pupils who like the approach. Other pupils who dislike that approach will be put off. For every pupil to experience success, varied tasks have to be chosen to meet their individual abilities and interests. This is particularly true of less-able pupils.

3 Different stages in the learning process require different strategies:
Stage 1 Introducing a topic The work should be an investigation, a period of discovery of new knowledge.
Stage 2 Consolidation This is a time for practice and consolidation, for the development of proficiency with the newly acquired skills.
These different learning stages call for different kinds of teaching activities. For many topics a variety of activities is possible. Results can be discovered experimentally or theoretically. Be aware when giving the pupils experiments or practical activities that their skills with measuring instruments may not be great, and this can lead to unexpected results. Even with older children, rounding errors can make an activity seem unsuccessful.

3 Examples of variety

Every topic in maths could be taught in a variety of ways. 'Multi Modal Maths' is a rather grand-sounding phrase to emphasise the fact that when we teach maths we should teach in a variety of ways. Derek Haylock has suggested the 'Think Board' as an illustration of the different ways we can introduce and develop topics.

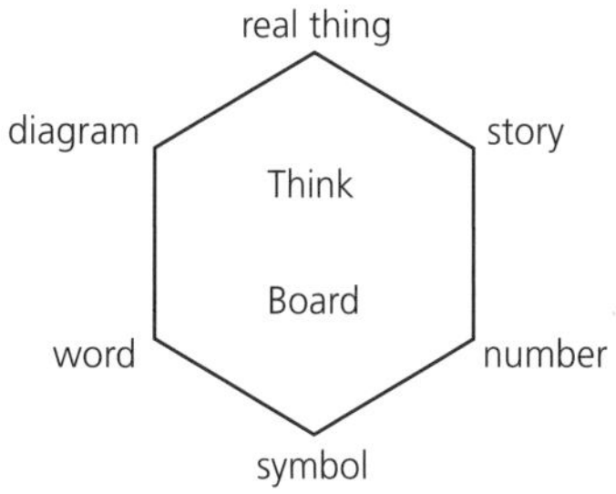

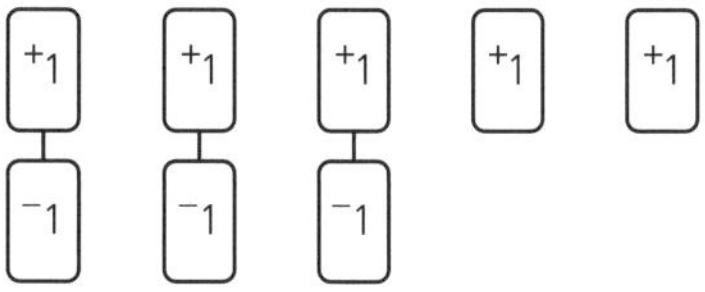

Teaching can be done through the use of:

- *Real things* (such as the squares and longs in arithmetic),
- *Diagrams* (such as pictures used in geometry),
- *Stories* (another name for word problems),
- *Words,*
- *Numbers,*
- *Symbols.*

As an example, consider the addition of integers (See Section B20):

Real thing

Represent $^{-}3$ by three '$^{-}1$' cards, and $^{+}5$ by five '$^{+}1$' cards.
We can show $^{-}3 + {}^{+}5 = {}^{+}2$ thus:
We can pair a '$^{+}1$' card with a '$^{-}1$' card, and they cancel each other out. We can do this three times.
Three pairs of cards cancel out, leaving two '$^{+}1$' cards.

Diagram

A diagram was used above to show the *Real thing* representation. One could have been used to show a number line. Pupils should meet a variety of representations as they learn maths — they do not need to meet all of them for each topic, but they should be exposed to a variety of them often.

Story

Playing away from home our team lost 0 – 3. In the return match at home we won 5 – 0. What was the goal difference between the two teams?
We had three goals scored against us, but scored five goals ourselves. They scored 3, we scored 5. The difference is 2.

Words

What is negative three plus five?

Numbers

Work out $^{-}3 + {}^{+}5$.
Think about a number line: go back three steps, and forward five steps. You have gone forward two steps altogether.

$$^{-}3 + {}^{+}5 = {}^{+}2$$

Symbols

$^{-}a + {}^{+}b = b - a$

Games and puzzles

Games and puzzles can also be used. **Snap** is a game to use with addition of integers.

Snap

(For 2, 3 or 4 players. A pack of 36 cards with 4 cards each numbered $^{-}4$ to 4.)

1. Deal all the cards face down, with an equal number of cards for each player.
2. All players turn their top cards over together.
3. If **the sum of any two cards is $^{-}$3,** the first player to call 'snap' *and* to name the two numbers is the winner and keeps those cards. If no two cards add to $^{-}3$ then no one is the winner and the cards are left on the table.
4. The next cards from the piles are turned over and the procedure is repeated.
5. The winner is the one with the most cards in their hand at the end of a certain time or when the other players run out of cards.

[From *Springboard Mathematics* by R. Palmer (Nelson, Australia)]

4 Tests

Tests can be made more entertaining if the questions are linked together to form a story. Here is an example:

Safari in Africa

As a reward for good work in maths, the school awarded Shupiwe, aged 12, Thandi, aged 14, and Ahmad, aged 16, a trip to the Luangwa Valley Game Park in Zambia. My wife and I took them there.

1 The cost of the holiday was $1400 for adults and $1200 for children under 16. What was the total cost of the trip?
2 We left Dakar at 22:45 and the journey took 9.5 hours. The time in Zambia is 2 hours ahead of the time in Dakar. At what time did we reach Zambia?
3 We spent a day in the capital, Lusaka. In a curio shop, a carving was on sale for 40 000 Kwacha. If $1 = K4000, what was the price in dollars of the carving?
4 Ahmad is very good at bargaining. He managed to get the price reduced by 30%. How much did he pay for the carving (in Kwacha)?
5 The next day we left Lusaka for the Game Park some 600 km away by air. The flight took 1.5 hours. What was the plane's average speed?
6 On arriving at the airport, we were picked up by a Land Rover to travel to the Game Lodge. On the way, we were stopped by a herd of elephant eating leaves next to the road. An adult elephant can eat half a tonne of food per day, and a young elephant eats one quarter of a tonne. In the herd, there were 5 adults and 8 youngsters. How much food would the herd eat each day?
7 Finally we reached the camp at 6.30 p.m., after dark. The following day we had to get up very early, as the animals are easiest to see when it is cool before the sun gets too high in the sky. At 5 a.m. we set off across the park with a game warden. We came across some zebra and waterbuck. Thandi announced that the animals had a total of 224 legs. How many animals did we see?
8 Under some trees, there was a group of 36 animals. They were zebras and ostriches. These animals had 96 legs altogether. How many were zebras?
9 Suddenly, out of the trees came a cheetah chasing the animals, but it didn't manage to catch any of them. A cheetah can run at up to 120 km/h but can only keep up this speed for one minute. How far can a cheetah run in that time?
10 In the game park, the ranger told us that 20 years ago there were 60 000 elephants but, because of poaching, the number had dropped by 70%. How many elephants are left now?
11 After five days, we flew home. The temperature when we left Zambia was 29°C. After the aircraft took off, we were told that the temperature outside was ⁻4°C. What was the change in temperature from on the ground to in the air?

Activity

1 Write an interesting story test for a class.

2 Choose a couple of topics and find ways of teaching that involve as many of the methods on the Think Board as you can.

A7 Class organisation

1 Introduction

Just as there should be a variety of approaches to teaching, a class can be organised in various ways to be taught:

- as a whole class
- in groups
- individualised learning.

Whole-class lessons

The class is taught as one group. Often the teacher will explain the meaning of a new concept or how to carry out a new skill. Afterwards, the children solve some problems, often on their own.

Lessons like this are popular with teachers because:

- preparation by the teacher is simplified
- each lesson is complete, with a presentation, explanation and conclusion
- the syllabus can be easily broken down into a sequence of lessons
- this was how most teachers learnt
- the teacher is in control of the lesson.

However, the method has disadvantages:

- it does not take account of different speeds of learning for different pupils
- the more able pupils may be bored while the less able are struggling to keep up
- what was thought by the teacher to be fully understood by the class, may not have been understood.

Group work

Pupils work in groups on a task. They help each other. Often group work involves physical activities, such as measuring distances, making models, and so on. The advantages of group work include:

- physical activities which produce better understanding of concepts
- creation of better social interaction among the pupils
- better pupil participation and attention through physical activities.

Disadvantages include:

- extra preparation, as there is often a need to organise equipment
- it can be more difficult to control the class, when different groups are doing different things
- some members of groups may do very little.

> ***Activity***
>
> Consider the advantages and disadvantages of whole-class teaching and group work. Many teachers use a mixture of both methods.
>
> Consider how this could be organised in your school.

Individualised learning

This can be interpreted in two ways. Pupils learn on their own, often from a book, on work set specifically for them. This is normally only used when there is no teacher. Alternatively, if there is sufficient staff, an individual pupil may be taught on a one-to-one basis, to help to overcome a particular problem.

2 Exposition techniques

We need to use a variety of approaches to teaching, because of different abilities, backgrounds, and attitudes amongst the learners.

Different kinds of learning are appropriate at different stages in the learning process. The introduction of a topic suggests investigation and a period of discovery of new knowledge.

Later, there is a need for practice and consolidation, for the development of proficiency with the newly acquired skills.

For many topics a variety of approaches are possible. Results can be discovered experimentally (beware of experimental and rounding errors) or theoretically.

3 Drill

Drill can be used for the development of computational skills, but must always be accompanied by flexible thinking. Pupils must be encouraged to think about what they are doing as they apply the methods they have learnt.

> **Example** 649 + 345 + 655
>
> This is much more quickly done if the pupil notices that 345 + 655 = 1000. Working out 649 + 345 = 994 first takes much longer.
>
> So 649 + 345 + 655 = 1649

The calculation $\frac{1}{2} + \frac{1}{2} - 1$ can be done very quickly if the pupil notices that $\frac{1}{2} + \frac{1}{2} = 1$ and realises that the answer is $1 - 1 = 0$.

Pupils taught to apply 'the fraction method', without thinking would make the calculation more difficult for themselves, and would be more likely to make mistakes.

4 Practice

There is no doubt that **practice** is very important in the maths lesson. It helps pupils to remember what they are doing, makes them more accurate, makes them more efficient at tasks, and makes them more confident.

It is part of the lesson and must be planned. The exercises should be made up with care. Here are some suggested **principles of practice**.

- Practice should be done only when pupils understand what they are practising.
- Practice should be brief, focused, and at regular intervals.
- Practice should emphasise principles rather than tricks and short cuts.
- Practice should involve varied activities, such as games, contests, puzzles, timed exercises, mental computations, group activities, oral and written exercises.
- Practice should never be given as a punishment. (Maths should never be considered as a punishment.)

Children need to learn how to practise — this will be very useful when they come to revise for examinations later. The intention always should be to benefit the children.

Activity

Prepare some warm-up exercises that could be used to give pupils practice in some topics of your own choice.

5 Short warm-up exercises

These are an excellent way to start a lesson. They bring the attention of the class to thinking about maths, as well as improving their ability to calculate fast and accurately.

6 Schemes of work

A school has to take the maths syllabus and develop a teaching programme which states which topics are to be taught, and when. Note that the order in which topics are listed in the Curriculum for a particular year does not mean that they must be taught in that order. If that were the case, all the arithmetic would be taught early in the year, followed by geometry and then statistics, for this is the order in which topics are usually listed in syllabuses. Each topic would be met annually. By the time the topic comes up again, a whole year would have gone by.

A better scheme of work allows pupils do some work each term in each of the main areas of the syllabus. This ensures variety in the work, and pupils are enabled to remember what they have learnt because they regularly revisit each topic.

A8 Solving word problems

1 Introduction

Mathematical word problems cause many pupils difficulties. They often find it hard to choose the correct operation to use. Word problems are even more difficult when they are being asked in a second language.

2 The four steps to solve word problems

There are several stages to go through when solving a word problem.

Stage 1 The pupil has to understand the problem. This means that he or she has to understand the words and the meaning of the problem as a whole.
For example, consider the problem, *What is the product of 5 and 18?* A pupil who does not know what *product* means cannot answer it.

Stage 2 The pupil has to **translate** the problem into **mathematical language** correctly. This means that he or she has to choose the correct operation(s) for solving the problem.
For example, consider the problem, *I have six toys. This is three more than my brother. How many toys does he have?* Many pupils see the word *more*, choose addition and apply it to the numbers they can see. Hence they work out $6 + 3$.

Stage 3 The pupil has to **calculate** correctly, using the operation(s) that he or she has chosen.
For example to solve the problem, *How many players are there in 15 football teams?* a pupil may answer $11 \times 15 = 155$, which is incorrect (as the pupil has forgotten to carry 1).

Stage 4 The pupil gives the answer in the **correct form**. If the answer is expressed in the wrong units, the answer is wrong.
For example, referring to the second example, $6 - 3 = 3$ is not sufficient. The answer should be stated as '3 toys'.

To solve any problem, the pupil must go through all these stages correctly.

Example 1

It takes me 40 minutes to walk to school from my house. School starts at 7.30 a.m. At what time must I leave my house?

1 **Understanding** To solve this problem, the pupil has first to understand the words and what the problem is asking. The phrase *it takes me* might puzzle some pupils — where is he being taken?

2 **Translation** If the pupil understands the question, he has to translate the problem into mathematics. The time to be found is 40 minutes before 7.30 a.m. Subtraction is involved.

3 **Calculating** It is not straightforward to subtract 40 minutes from 7.30. Many pupils will forget there are 60 minutes in an hour and treat the figures as if they are decimals (7.30 –.40 = 6.90, 6.90 a.m.?). Some will make the mistake of simply subtracting the smaller number from the larger number and write 7.30 – 40 = 7.10. The correct solution requires 'borrowing' 60 minutes from the hours (7.30 – 40 $\Rightarrow$ 6.90 – 40 = 6.50), giving 6.50 a.m.

4 **Correct form** The correct form of the answer in this case is to state a time. It can be given as 6.50 a.m. or 06:50.

Activity

Look at word problems in a chapter from your textbook. For the questions:

(i) decide whether your pupils can understand all the words
(ii) decide whether your pupils can understand what is being asked
(iii) discuss what difficulties pupils are likely have choosing the correct operation(s) and doing the calculations
(iv) discuss whether the easier questions are at the beginning of the exercise and the more difficult questions are at the end.

Example 2

Find the area of a corridor that is 20 m long and 80 cm wide.

1 **Understanding** The pupil has to understand what *area* means. He does not really have to understand the meaning of *corridor*.

2 **Translation** To translate the problem into mathematics, the pupil has to realise that area = length × breadth and so multiplication is involved.

3 **Calculating** The pupil has to multiply 20 m by 80 cm correctly. He or she has to realise the units are given in different forms and must be changed so that they are the same. Simply multiplying 20 × 80 to get 1600 would lead the pupil to write 1600 m^2 or 1600 cm^2, both of which are incorrect. The correct calculation is 20 m × .8 m = 16 m^2. (2000 cm × 80 cm = 160 000 cm^2 is also correct, but not so easily visualised.)

4 **Correct form** Unless the pupil is careful to set up the multiplication correctly, it would be easy to muddle the units involved.

3 Key words

More Many pupils think that when a problem contains the word *more*, the problem should be solved by adding numbers. The strategy sometimes works, but it often leads to the choice of the wrong operation.

For example, *I have 6 oranges; my brother has 2 more. How many oranges does he have?*

The pupil sees *more, 6*, and *2* and decides the problem is solved by adding 6 and 2. The strategy works in this case.

However, if the problem were *I have 6 oranges. This is two more than my brother. How many oranges does he have?*, the addition strategy would be wrong. This problem is solved by working out 6 – 2. The meaning of the statements has to be interpreted before operations can be decided on.

Some pupils have a different strategy. They think that because the word *more* is used, then the answer is the larger of the numbers in the problem. In both of these problems, a pupil who thought like that would answer 6.

Less The word *less* causes similar problems. Some pupils think it is an instruction to subtract. They think that the answer to the problem *I have 6 oranges. This is two less than my brother has. How many oranges does he have?* is 6 – 2 = 4. Again, correct interpretation leads to the solution, 6 + 2 = 8.

Altogether Many pupils think *altogether* is an instruction to add. This leads to the correct answer in *There are 12 girls and 16 boys in my class. How many pupils are there altogether?* However, look at the problem *There are 28 pupils altogether in my class. There are 16 boys. How many are girls?* Adding 28 and 16 because the word *altogether* is in the question leads to a wrong answer. (In this case *altogether* implies that the number of boys and the number of girls has already been added.)

Activity

1 Write down ten words of comparison that are often found in word problems.

2 Write some word problems that includes the words:

- *longer* where the solution requires subtraction
- *shorter* where the solution requires subtraction
- *altogether* where the solution requires subtraction
- *less* where the solution requires addition.

Other words of comparison can also cause difficulties. These words include *longer*, *shorter*, *higher*, *narrower*, *thinner*, *thicker*, and many more.

4 Teaching your pupils to solve word problems

Pupils need help in learning how to solve word problems. In computation problems (such as *What is 25% of 200?*), the question tells the pupil what to do. For any word problem interpretation is the key to finding the solution. Help your pupils to follow the steps outlined in part 2 above.

1 Ask your pupils to read the problem.
2 Ask them to explain to you what is being asked. (This will help you to know if they have understood the problem.) Are there any words they find difficult?
3 Do they understand what is *known* and what is *not known* in the problem?
4 Can they choose the correct operation(s) to find the solution?
5 What sort of problem is it?
Is it a one step problem? *I have five apples. My sister has two fewer than me. How many apples does she have?*

> ***Activity***
>
> Choose a topic from the syllabus. Prepare a set of problems that your pupils can use to practise problem solving. Make sure that they are not too difficult.
>
> Discuss how the problems can be modelled in class.

Is it a two step problem? *A girl reads for 20 hours each week. If she reads for 10 hours at the weekend, how many hours, on average, does she read each weekday?*

6 Can they carry out the calculation in the correct way?

7 Does the answer make sense? Is it in the correct units? It is a reasonable and sensible answer? They should get into the habit of checking their answer. This is an important matter. Pupils should be encouraged to check that their answers are sensible ones.

5 Modelling

Pupils can be helped to find a correct method to solve a problem by modelling similar situations.

For example, *'There are 250 boys and 280 girls in a school. How many more girls are there than boys?'*.

This could be illustrated with, *'There are 10 boys and 12 girls in this class. How many more girls are there than boys?'*. Ask 10 boys and 12 girls to stand up, pair them one boy to one girl, and it is clear that we need to find the **difference** of the two numbers, and not the sum.

Using small numbers should help to make it easier for the pupils to see how to do a problem involving large numbers.

A9 The language of mathematics

1 Introduction

Skemp suggested that, in addition to relational understanding, pupils have to acquire what he called 'formal' understanding of the language of maths. Maths is not a language in the way that English or Kiswahili are languages, but it has its own vocabulary, notation, rules and conventions. In a piece of English text, there are often clues that help with finding the meaning of new words. In a piece in a maths textbook, there are often very few clues. This makes maths much harder to read. How can you help your pupils to learn this language?

> ***Activity***
>
> Read a chapter in your school textbook.
>
> Make a list of the words used in the book that will be unfamiliar to pupils.
>
> Discuss how you would teach them to your pupils.

2 Learning mathematical vocabulary

Pupils have to learn the meaning of many words in maths. The highly specialised technical vocabulary used in maths can deter people from enjoying maths. Words such as *integer*, *index*, *congruent*, *multiple* are used in class, and rarely in everyday life, so the pupils do not become familiar with them. Too many unknown words used in textbooks can be intimidating. What can you do the help your pupils learn this vocabulary?

3 Meanings of words for things you can see

You can teach your pupils these words by showing them examples.

Here is one way to teach what a *rectangle* is.

- Give each pupil a rectangular card or piece of paper.
- Get them to play with them, and make patterns with them, measure angles and lengths, etc. The pupils take part actively in this and discover properties of rectangles for themselves.
- Encourage your pupils to talk about what they are doing, and what they find out, in their own words. Can they see these shapes in the classroom? Do not emphasise correct English too much. If the pupils feel that their English is always going to be corrected, then they will feel less confident about speaking freely.
- At the end, sum up what the properties of a rectangle are. Use the mathematical words often so that the pupils become familiar with them quickly.

4 Meanings of words for things you cannot see

You can show your pupils a rectangle, but there are many mathematical concepts that you cannot show them, such as, *square root*. However, you can show them examples:

4 squared is 16, so we say that 4 is the *square root* of 16.

Avoid giving formal definitions — examples are sufficient. Use the new word often.

> ### Activity
>
> 1 Make a list of English words that have one meaning in maths and another meaning in everyday life. State what the different meanings are. Discuss how you would teach the two meanings.
>
> 2 How would you teach the mathematical meanings of these words: *capacity*, *volume*, *isosceles triangle*, *pyramid*, *cone*, *area*, *represent* and *stand for*?

5 Words that have different meanings in mathematics

Root is a word that has a special meaning in maths. In everyday use, *root* means something quite different.

Axes is the plural of *axis,* a mathematical word for a line. It is also the plural of *axe*, used for cutting down trees.

Other words that have a special meaning in maths and a different meaning in everyday usage include *product* and *term.*

Represent and *stand for* are puzzling. Consider sentences like 'n represents the length of the line in centimetres.' (Compare this with 'The High Commissioner represents his country.') or 'n stands for the number of sweets.' (When it gets tired, does it sit down?). Can you explain these words in simple terms?

> ### Activity
>
> Consider these questions:
>
> - Is a square a rectangle?
> - Is a square a rhombus?
> - Is a parallelogram a trapezium?

6 Words in mathematics have a precise meaning

A square is a quadrilateral with equal sides and equal angles. A rectangle is a quadrilateral whose opposite sides are equal and whose angles are right angles. Is a square a rectangle? Many pupils think that because the words are different the answer is 'no'. Is that correct?

7 Symbols used in mathematics

Many symbols are used in maths. The equals sign (=) is one of the most important. Even teachers often use it incorrectly.

Children first use the equals sign in sentences such as '$4 + 2 = 6$'. '=' indicates the result of an operation; it can be stated as 'the result of adding 4 and 2 <u>is</u> 6' or '4 + 2 <u>makes</u> 6.'

Children can be puzzled by the sentence '$6 = 4 + 2$'. It seems to be the wrong way round, for we do not say '6 makes 4 + 2.'

However, the sentence is correct, mathematically. The '=' sign means that the expressions on each side of the sign are the same. It does not matter which way round they are written.

An equation such as $5x + 7 = 22$ fits the pupil's earlier perception of the meaning of '='. When the sentence is reversed, as in $22 = 5x + 7$, some children can have problems, but they should be reassured that the sentences are identical.

Moreover '=' is used in other ways. It is used in *identities*, such as $2(x + 5) = 2x + 10$, in *functions*, such as $y = 3x + 4$, and in *formulae* such as $V = l\ b\ h$.

> ***Activity***
>
> Make a collection of examples of sentences in which '=' is used in the wrong way.

8 What can the teacher do to help?

- Include sentences like 7 = 2 + 5 early in their number work.
- When teaching equations include problems like 'Solve 25 = $4x$ – 7.'
- Do not use '=' as a shorthand for 'represents' or 'take.' For example, '1 cm = 10 km' on a map, and '4 men = 52 hours' are nonsense. Only use '=' when the two expressions are equal.

9 Learning the rules and conventions of mathematics

Order of operations

Say to your pupils *'Work out 4 add 3, multiplied by 5.'* They will probably answer 35. However, 4 + 3 × 5 (which seems to be the same statement written in symbols) has a different answer.

$$4 + 3 \times 5 = 4 + 15 = 19$$

because the agreed rule in the language of maths is to do multiplication before addition. The statement in words should have been written with brackets as

$$(4 + 3) \times 5 = 7 \times 5 = 35$$

to show that the addition should be done first.

To avoid confusion everyone has agreed to carry out operations in this order:

brackets first	then	**multiplication or division**	then	**addition or subtraction**

This rule is often called **BODMAS** (**b**rackets, **o**f, **d**ivision, **m**ultiplication, **a**ddition, **s**ubtraction.) S is merely put after A in order to make the letters into a word that is easy to say. (Try saying BODMSA!)

A10 Teaching mathematics in the English language

> ***Activity***
>
> Conveying ideas can be affected by how easy it is to express them. Are there words used in maths that are difficult to translate into your local language?

1 Introduction

You will probably be teaching maths to pupils whose first language is not English. That means that your pupils will need to speak about maths, to listen to a teacher talking about maths, and to write about maths in a foreign language. What can you as a teacher do to help them do these things?

2 Counting words

Counting in English can be tricky. The words for numbers like 61 are straightforward: 'sixty-one', meaning '6 tens and 1'. In this case, in both the words and the figures, the tens number comes first.

However, the numbers between 10 and 20 are different. For example, 'thirteen' in figures is '13' and 'thirteen' clearly indicates 'three ten.' However, the three is said before the ten, while in the figures 13, the ten is before the 3, This can cause confusion and it is not common to all languages.

In Nsenga, a language of eastern Zambia, 3 is 'vitatu', 10 is 'kumi' and 13 is 'kumi na vitatu' which means ten and three. The words and symbols are in the same order.

> ***Activity***
>
> Investigate the number words in your area.
>
> Write down the words for 12, 21, 35 and 58. Are they written in the same order as the symbols that we use?

The English words *eleven* and *twelve* are even more difficult and seem to have no connection to '10 and 1' and '10 and 2,' The origin of these words is in counting on our hands. Eleven is derived from 'one left over' (*one leave*), twelve from 'two left over' (*two leave*), when counting 11 and 12 things using our ten fingers. Some people have suggested that the English language should be changed for the words from 11 to 19!

3 Learning in English

Pupils having to learn maths in a second language can have difficulties and make errors both when speaking and writing in the second language. Here are some common questions that arise on the subject of correcting English.

> ***Activity***
>
> Discuss these questions and decide what you think your school policy should be. Should the policy be different for different grades?

- Should the teacher correct the use of English in their written work? Clearly, she should, but how far should she go? Should she mark all errors? Too much correcting can be discouraging. It is better to concentrate on the major errors in a piece of work.
- How much should the teacher correct pupils spoken English in class? Is correction going to result in embarrassment for the pupil? Correct use of words is very important, but what about errors that are only grammatical?
- By how much should teachers reduce marks they give if the quality of English is not very high?

4 Suggestions for teaching

In addition to correcting errors, what can teachers do to help their pupils understand the maths better as they learn in English? Experience suggests that teachers can take steps to help their pupils improve their understanding of what is going on in the maths lesson. The pupils must have plenty of opportunity to practise talking in English about the maths that they are learning. People are usually good at what they do often. 'Practice makes perfect.'

Ideally, you should:

- use simple constructions and avoid long sentences
- set a good example and be error-free, precise, fluent, and clear
- introduce only a few new words at a time and use these new words often
- use the active tense (*'Joseph gave me $10.'*) rather than the passive tense (*'I was given $10 by Joseph.'*)
- avoid using the conditional tense (*if, unless*)
- use real examples and illustrations to develop new concepts
- make your pupils use English a lot, by asking them many questions as you teach — the more they speak in English the better they will be
- ask questions to individual pupils so that each class member is encouraged to speak on his/her own — do not use chorus answers
- do not correct the pupils' English too much. It can be embarrassing and discouraging. If meanings are wrong correct gently — but do not worry too much about incorrect grammar. They are only Primary school children, and you want to build up confidence at this stage.

Activity

Write the following sentences in simpler English.

1 It is absurd to say that more than 100% of a class have passed a test. However, in business transactions there are situations where percentages greater than 100% arise.

2 Shupiwe saved $5 on a pair of slippers she purchased at a clearance sale. This represented 20% of the original price. What was the original price of the pair of slippers?

A11 Applying mathematics

1 Introduction

Maths has an important place in the curriculum because it is a skill used in everyday life. Pupils will use and apply what they learn in maths in their daily lives. This does not mean that they first learn the maths and then apply it. Daily life uses the application of maths, school learning helps pupils to gain a deeper understanding. This also means that in order to be skilful at using and applying maths, pupils should learn maths in the context of solving problems.

Other units have suggested that topics should be introduced by trying to get pupils to construct knowledge for themselves. To do this, they will be using maths. Using the example from an earlier chapter, the teacher does not tell the pupils the formula for the area of a rectangle, but asks the right questions to enable them to find it for themselves. The pupils apply what they know to extend their knowledge, with the help of the teacher.

2 Applying mathematics

Children can apply maths in several situations. They can apply it when they are undertaking **practical activities**, when they are **solving problems**, and when they are carrying out **investigations**. Here are some examples of each of these.

Practical activities Examples include the following sorts of activities:

- finding the number of pencils two people have if one has 3, and the other has 5
- building models — lots of mathematical opportunities in preparing nets, measuring, etc.
- comparing and estimating sizes, weights, numbers, etc.
- making symmetrical paper darts or paper hats
- measuring quantities when cooking
- sorting objects according to colour, shape, use, etc.
- using objects to work out 12 times 7.

Solving problems These can be theoretical or practical problems.

- If 8 pupils have 3 pencils between them, how many more pencils are needed for all to have one?
- Is there a rule that will tell you prime numbers?
- Read and follow instructions for a game.
- Arrange the chairs in the school hall.
- Make a bed for vegetables in the school garden.
- Decide how many trees should be in an orchard.
- Make a map of class seating using coordinates.
- Decide on 'best buys' in the supermarket.

Investigations Again, these can be 'paper problems' or practical activities.

- Find numbers that are multiples of 3 and 4.
- Is it true that rectangles with the same perimeter have the same area?
- What is the most likely score when a die is rolled?
- Magic squares.
- Why is a given amount of water higher in a thin jar than fat one?
- $5^2 - 4^2$: is there a quick way to work out the difference between two squares?
- Cereal box: what should you do to the length, breadth and depth of a cuboid in order to increase the volume by 20%?
- Fixed perimeters: what shapes can you make that have a perimeter of 24 cm?
- How far up can members of this class leap from a standing position?
- Is the length of a person's foot related to his height?
- The ancient Egyptians believed that the length of your foot is the same as the length of your forearm. How true is this belief?
- Investigate traditional methods of measuring length, time, capacity, etc. in your area. Compare with measuring using the metric system.
- Investigate traditional ways of counting with the base ten system.

3 Pupils deciding, telling, reasoning

The activities above all require pupils to **decide** what to do next as they work. For example, some pupils are making the net for a box. They know how big they want the box to be. They have to decide the shape of the net before cutting the paper.

Pupils should be required to **tell** others about what they have done. In life, we have to tell others about what we have done. Having to explain something to others is a good way of developing the ability to express yourself, it helps you to think for yourself, and it tests whether you really understand what you are trying to explain. Pupils also learn to learn from what others are saying when they talk, which is an important skill. It also improves their ability to read maths. If pupils carry out surveys, they can present the information they have discovered as tables or graphs.

Pupils need to learn to **reason**. The activities above should encourage pupils to predict, generalise, discuss and argue (in the best sense of the word) about what they have found out.

4 Helping pupils with projects

Pupils who are used to being *told* how to find the answer will need gentle assistance with projects. Here are some suggestions of what can be done to help them.

- When presented with problems, pupils need time to think about them. Pupils learn to think by thinking.
- When pupils need help, if the teacher just tells the pupils the answer he or she will have deprived them of the pleasure of solving the problem. Intervene discreetly and help them in such a way that they still feel that the problem is their own. Teachers should not intervene with the answer (i.e. they must not steal the problem). Less-able pupils need more help, but they can still succeed.
- Allow group work to encourage cooperative learning.
- Encourage pupils to communicate results through reports, displays and a class presentation. Writing up ideas clarifies them and ensures meaningful learning.
- Try not to be prescriptive and be willing to accept unorthodox methods.

Through project work, pupils:

- develop a greater sense of achievement, and are more willing to explore and discuss methods
- are more willing to talk about their **own methods**
- cease to expect problems to have instant solutions
- develop increased confidence in their problem-solving ability, and are more willing to tackle problems that they previously thought too hard.

5 How should projects be assessed?

Marks should be given for the method used, for whether the maths is correct, the quality of the communication of results and for the appearance of what is produced.

Here is a possible marking schedule:

• method used	6 marks
• correctness of maths	6 marks
• quality of explanation	4 marks
• neatness	4 marks
• initiative	bonus of up to 5 (as a reward and incentive for pupils who take the project further than expected)

Activity

Develop a programme for trying out some projects with your pupils. Begin with small projects that take a short time.

A12 Helping slow learners

1 Introduction

Many pupils have difficulties in learning maths. Teachers need to be sensitive to their difficulties and help them to overcome them.

2 Some causes of difficulties in learning mathematics

There are several reasons why maths is difficult for many people.

- **Problem solving** In maths, pupils are not usually asked to recall facts. The basic facts are assumed and are used to solve problems. For example, pupils may not be asked to state the formula for the circumference of a circle. Instead they are given the formula, told the radius and asked to work out the circumference. They are assumed to know the meaning of the terms and the relationship between them. They have to use this knowledge to solve the problem.
  Pupils are expected to solve problems presented in everyday language. Learners with special needs often find the choosing of appropriate strategies difficult.
- **Language** Maths is highly abstract and has a special, very precise language of its own. The language gives maths its power, but also makes it more difficult. A slight change in the symbolism leads to a major change in meaning. Compare these problems, each of which differs marginally from its neighbour.

  $$\begin{array}{r}14\\+\ 6\\\hline\end{array}\qquad\begin{array}{r}41\\+\ 6\\\hline\end{array}\qquad\begin{array}{r}14\\\times\ 6\\\hline\end{array}\qquad\begin{array}{r}41\\\times\ 6\\\hline\end{array}$$

  For some pupils distinguishing between these is very difficult indeed.
- **Developmental nature of maths** Maths builds on itself. What we learn depends very much on the **prerequisites** having been learnt. Success at any stage has to be preceded by success at all the earlier stages. If the pupils do not remember the previous work, they cannot advance. For example, basic arithmetical calculations depend on knowing both the basic number facts and the algorithms. A mastery of arithmetical calculations must occur before competency in higher order arithmetic work can be achieved.

3 Assessing difficulties in mathematics

A doctor undertakes tests to ascertain the causes of a patient's illness. A teacher undertakes tests to ascertain the causes of a pupil's difficulties. It is important that teachers learn how to assess the difficulties that their pupils have. This can be done in several ways.

- **Observe** pupils at work. Note who does their homework, note how often each pupil needs help with their work, and so on. Be sensitive to events in the classroom.
- **Error analysis** helps teachers to understand the causes of the pupils' errors. Pupils make errors for many reasons.
  Different types of error:
  (i) Incorrect operation The wrong operation is used to solve a problem. Give practice with problems where pupils decide which operation to use, but do not work out to a numerical 'answer'.
  (ii) Incorrect number fact The number fact is wrong, e.g. $6 \times 9 = 52$. One solution is to use flash cards and other games to develop speedy recall of basic number facts.
  (iii) Inaccurate algorithm A pupil follows the wrong procedure in a computation. A possible solution is for the teacher to build up the algorithm carefully from simple examples, progressively working from easier examples (e.g. no carrying or borrowing) to harder ones. This is where task analysis is very helpful and effective.
  (iv) Carelessness There is not much we can do about this — or is there? Can we train pupils to check their answers? Can they use estimates to check for silly answers? It is difficult to eliminate carelessness altogether, though estimates and checking can reduce it considerably.

Several steps should be followed when undertaking *error analysis*:

- get a sample of the pupil's work, e.g. homework, exercise book, tests, and so on
- mark the sample, paying great attention to the errors made
- categorise the errors
- look for repeated errors and decide whether the errors are in some way systematic
- give the pupil work which will help to eliminate the error.

> ### Activity
> Obtain the exercise books of some weaker pupils and analyse the errors that they have made in their work. Decide on a plan (or plans) that will help the pupils.

4 Error patterns in computational work

The teacher should analyse pupils' answers to find causes of errors. For example, look at these errors. Each error has an explanation next to it. Knowing this, a teacher can correct the error.

(a) **Addition**

$$\begin{array}{r} 74 \\ +\ 56 \\ \hline 1210 \\ \hline \end{array}$$

The pupil adds and writes down the results without carrying.

(b) **Subtraction**

$$\begin{array}{r} 524 \\ +\ 298 \\ \hline 374 \\ \hline \end{array}$$

In each column the pupil subtracts the smaller number from the larger number.

(c) **Multiplication**

433 × 226 878	In each column, the numbers are multiplied together. When the result is more than 10, 1 is carried to the next column.

5 Task analysis

The teacher should consider what the prerequisites are for the topic causing difficulty. It may well be the problem is caused by one of them.

A checklist on the topic being taught can be very useful. For example, if the topic is 'telling the time', the list might include:

- can identify time to the minute by saying '... minutes after the hour'
- can tell time by fractions (half an hour, quarter of an hour).

6 Probe

In order to decide which of the prerequisites is causing difficulty, the child can be asked to do a test that contains questions on each prerequisite. This test is called a **probe**.

Here is a probe for multiplying a two-digit number by a one-digit number.

1 (a) draw 6 yams (b) draw 9 yams (c) draw 15 yams
(This tests number recognition.)

2 Arrange these numbers from the largest to the smallest:
(This tests knowledge of place value.)
42, 18, 65, 9, 37, 60, 43

	(a)	(b)	(c)	(d)	
3	2 + 5	5 + 7	8 + 9	35 + 9	(This tests knowledge of addition.)
4	2 × 3	5 × 1	8 × 4	6 × 9	(This tests knowledge of multiplication facts.)
5	14 × 3	23 × 4	41 × 4	87 × 9	(This tests knowledge of multiplication.)

Activity

Choose a topic that you are going to teach, decide on the prerequisite topics, and write a probe for your pupils to use.

7 Interview

Talking with the pupil helps the teacher to find out the processes the pupil uses in solving a problem. It helps the teacher to understand the way the pupil thinks. The interview can be conducted in several ways. One way is to ask the pupil to write out his/her answer to the problem. The answer is looked at and the pupil explains how he/she worked it out. Alternatively, the pupil is interviewed whilst doing the problem.

In the interview, ask the child to work aloud. Ask questions such as:

- *Can you work it out?* (Suggest the use of aids: fingers, counters or number line.)
- *Can you explain what to do?* (The answer to this question often reveals the point of confusion.)

If the pupil cannot answer, ask simpler questions, such as those in the probe.

The interview technique is open to error. Interviewing a child during an activity may influence how he/she does it. Afterwards the child may not remember what he/she did.

A13 Mathematical games

1 Introduction

Children enjoy and are motivated by games. They can be useful tools in the teaching of maths. If the environment in which pupils learn maths is good, then their attitude towards maths will be positive. Games help to promote a **good atmosphere**.

2 Games in the classroom

Games are helpful when teaching maths for the following reasons.

- They are **enjoyable**. They bring excitement and enthusiasm to lessons as pupils strive to do well and win. Repetitive exercises can become boring. 'Work' is much more interesting when success in the exercises is necessary for success in a game.
- Games make **revision** more interesting. When one of these games is played, the pupils may do a lot of calculating. They do not notice this. They are concentrating on the game. In a game of 'Maths Bingo', for example, players may work out about 15 problems.
- **Less-able pupils** who do not normally participate much in class become actively involved. Success in games often depends on luck, so they may find themselves defeating more able classmates. This brings satisfaction. Their self-confidence grows.
- Games can be valuable in improving pupils' '**mental maths**'. Exams emphasise working quickly. A pupil who calculates and reasons quickly has a big advantage over a slower pupil. Games encourage pupils to calculate quickly, as pupils enjoy the fast pace of a game.
- Games help children to understand mathematical concepts, develop mathematical skills, learn the vocabulary, notation and language of maths.
- Games also encourage the development of cooperative learning, promote discussion between teacher and pupil(s), and between pupils. They also contribute to the development of communication skills.
- Suggest variations to give games a longer life or simplify them for younger pupils. Encourage pupils to comment on the games and suggest changes. Encourage them to devise their own games.

3 Making your own mathematical games

Games can be made using many simple and cheap resources. The following materials are useful:

any paper unused on one side	old envelopes
wrapping paper	coloured paper
squared paper	old exercise books

If paper is not stiff enough, old envelopes can be used. Many of the games use **small cards** with numbers on them. These can be made from leftover card or old envelopes.

Playing pieces can be counters, buttons, coins, beads, and so on.

4 What makes a good game for the classroom?

A good game has clear **instructions**, simple easy-to-follow **rules**, and is quite **short** (10 to 15 minutes playing time). It is appropriate for a mixed-ability class, fun, and makes pupils want to play again.

5 Organising your games

- Store each game in its own container. Old large envelopes are very useful.
- Store all games together in a sturdy box.
- Label each game clearly.
- Dice games: use a shallow plastic tub (e.g. a margarine tub) to roll dice.
- Use 35 mm film containers, or other small containers with lids, to store small items.

6 Types of games

Maths games can be divided into two categories. Games suitable for use as **whole-class** games and games suitable for playing by **small groups** of pupils.

For each game, the **rules** of the game are explained and the **apparatus** required is stated. The apparatus used is freely available or easily made, such as dice, counters and number cards. Most games can be **modified**, and used at different levels. A game used to practise the addition of fractions, for example, can be modified quite easily to practise some other skill, such as substitution in formulae.

Games for the whole class (or a large group of pupils)

1 Flash cards The class is divided into two teams which compete against each other to answer problems on flash cards. The faster correct answer wins a point for that team. Questions for the flash cards will be based on the topic being studied.

Example of a flash card on basic multiplication facts:

Front	**Back**
5×8	$5 \times 8 \quad 40$

The question and answer are written on the back, so the teacher can read the card without turning it over.

2 Maths Bingo This is a game for a group of pupils with one leader. Participants choose six numbers from a list of about 15 numbers put on the board. (You can vary the numbers from game to game.) The leader then reads a succession of problems, whose answers are amongst the original list of numbers. The questions used will be based on the topic being studied. The first pupil whose six chosen numbers are all answers to questions given by the leader calls out 'Bingo!' The leader checks to see if the pupil's calculations are correct. If they are, he/she is declared the winner.

Example for Maths Bingo:
Write a list of numbers to choose from on the board.

25 30 35 36 40 42 45 48 49 54 56 60

Testing multiplication facts, the problems could be:

5×5 5×6 5×7 6×6 8×5 7×6 9×5 6×8 7×7 6×9 8×7 6×10

(The questions should be asked in a random order.)

Fractions problems associated with each of these numbers are:

$\frac{5}{8} \times 40$ $\frac{2}{3} \times 45$ $\frac{7}{10} \times 50$ $\frac{9}{10} \times 40$ $\frac{2}{3} \times 60$ $\frac{1}{4} \times 168$ $\frac{3}{4} \times 60$ $\frac{3}{5} \times 80$ $\frac{7}{10} \times 70$ $\frac{1}{2} \times 108$ $\frac{7}{10} \times 80$ $\frac{2}{3} \times 90$

Mental addition problems that could be used:

18 + 7	25 + 5	13 + 12	25 + 11	28 + 12	25 + 17
32 + 13	22 + 26	36 + 13	25 + 29	28 + 28	36 + 24

Variations The numbers, and the problems could be changed to suit the topic. (For example, make the additions all with or all without carrying.)

3 What's my number? The teacher chooses a number between 0 and 1000. The pupils have to find out what it is by asking questions, the answers to which are either 'Yes' or 'No'.

For example:

Is the number less than 50?
Is the number even?

If the pupils need more than 15 questions then they have lost. The pupils have to learn to think of the best strategies for getting the answer quickly. Questions early in the game such as '*Is it 213?*' are not very useful, although occasionally the pupils can be lucky and guess the right answer in this way. Questions such as '*Is the number more than 500?*' are much more effective in that the answer immediately eliminates half of the numbers.

Variations Make the number a decimal, a fraction or negative number. For younger pupils make the range 0 to 100.

4 Guess my rule The teacher makes up a rule for changing numbers that she tells no one. She then asks pupils to tell her a number. She writes the number on the board and then writes the result of applying the rule to the number. For example, suppose the rule is 'multiply by 3, then add 2' then, if a pupil chose the number 5, the teacher would write 17. Another pupil suggests a number. The teacher applies the rule. The pupils have to deduce the rule. The first pupil to correctly state the rule takes over, and invents his own rule.

5 I have ... who has ...? This game does not have a winner (or a loser), but it is fun and is good for settling a class down at the start of a lesson. Each pupil has a card with a number and an unrelated calculation on it. The first pupil says '*I have* (the number), *who has* (the calculation)*?*' For example if I say I have the card that says, '*I have 26, who has* 5×8*?*', everybody looks at his or her card. One person has the card with '*I have 40...*' and says, '*I have 40, who has...*' The cards are read around the class until they get back to the starter again. (All the cards in the set must be given out — some participants may need more than one card to use up the set. No two cards can have the same number on them.) The cards must be prepared in advance, so that you are sure the circle can be completed.

I have 40. Who has $7^2 - 6^2$?	I have 13. Who has $5^2 + 3^2$?	I have 34. Who has $4^2 + 1^2$?	I have 17. Who has $3^2 + 4^2$?
I have 25. Who has $10^2 - 6^2$?	I have 64. Who has $8^2 - 7^2$?	I have 15. Who has $7^2 + 5$?	I have 54. Who has $2^2 + 6^2$?

Here is an example of a set of eight cards using indices. For a class, you would need more cards.

Note that the solution to the last card is the number on the first card, thus making a complete circle, that can be started with any card.

Games for small groups of pupils

Some games are more suitable for small groups of pupils than for the whole class. However, a good way to show them to the whole class is to play them as whole-class games, with a few pupils playing and most watching. Thereafter, they are best played as small group activities.

1 Board Rush (For teams of about 6 pupils. This can also be played as a class game.)
The teacher writes about 6 numbers on the board. She asks one member of each team to stand by the board. She then asks these pupils a question whose answer is one of the numbers on the board. When a player has worked out the answer, he has to rush to the board and put a mark on the correct answer. The first one to do it correctly wins a point for his/her team. The team members take it in turns. The team with the most points wins.
It is best if the teacher prepares the questions in advance. She might choose the answers 45, 48, 54, 60, 64 and 72. The questions she then asks the players could be:

What is 18×3*?* *What is 80% of 60?* *What is* 30×1.5*?*
What is 8^2*?* *What is* $\frac{4}{5}$ *of 90?* *What is* $6^2 + 2$ *dozen?*

The questions can be changed to suit the topic that the teacher wants to focus on.

2 Card game For 2 or more players. You need a set of cards numbered ⁻15 to 15.

1 Shuffle the cards and deal three cards to each player.
2 Each player works out the sum of the largest and the smallest value card.
3 If the sum of the numbers is:
(i) positive, the player scores **2**
(ii) zero, the player scores **5**
(iii) a cube, the player scores **10**
(iv) anything else, the player scores **0**
4 The winner is the first player to reach a score of **50**.

[With thanks to Dave Womack]

3 The Beetle Game For 2 or more players. You need a set of cards numbered 1 to 100 in an envelope.

1 The object of the game is to draw a beetle with a body, a head, two antennae and six legs.
2 Players take it in turns to pick a number from the envelope.
3 If the number is **even,** the player can draw the **body**.
4 If the number is **odd,** the player can draw a **leg**.
5 If the number is a **square**, the player can draw the **head**.
6 If the number is **prime**, the player can draw an **antenna**.
7 A player must start with the body and must have a head before drawing an antenna.
8 The winner is the player who finishes their beetle first.

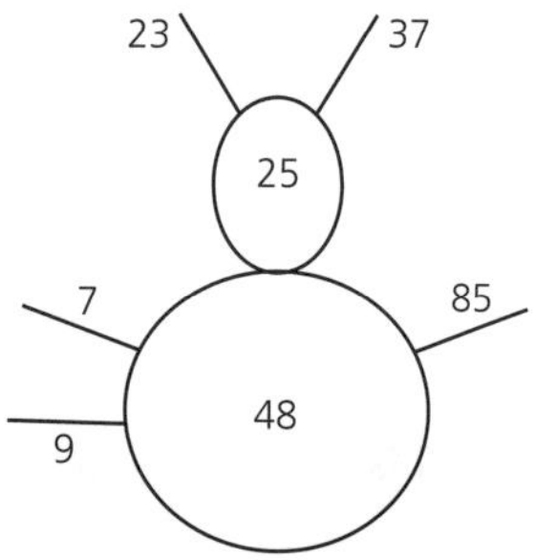

The diagram shows the result when a player picks these numbers:

53 (can draw nothing as body requires an even number)
48 (even, draws body)
7 (odd, draws first leg)
85 (odd, draws second leg)
66 (even, but not a square, cannot draw)
25 (square, draws head)
23 (prime, draws first antenna)
37 (prime, draws second antenna)
68 (even, so cannot draw)
9 (odd, draws third leg), and so on

Variation The Spider Game with eight legs. The rules for drawing the parts of the picture can be changed to multiples of numbers, and so on.

4 Snap For 2 players. A pack of special cards is required. The cards all have calculations on them. The answers for each group of 8 cards will be the same.

1 The cards are shuffled and dealt face down, with a pile for each player.
2 The first player places his card face up on the table. The second player puts his card face up next to the other player's card.

3 This continues until two successive cards have the same answer. The player who notices this and calls out 'SNAP' wins all the cards on the table.
4 The game ends when one player has no cards left. The other player is the winner.

Here is a possible pack of special cards for equivalent fractions:

$\frac{1}{2}$	$\frac{2}{4}$	$\frac{3}{6}$	$\frac{4}{8}$	$\frac{5}{10}$	$\frac{6}{12}$	$\frac{8}{16}$	$\frac{10}{20}$
$\frac{1}{3}$	$\frac{2}{6}$	$\frac{3}{9}$	$\frac{4}{12}$	$\frac{5}{15}$	$\frac{6}{18}$	$\frac{8}{24}$	$\frac{10}{30}$
$\frac{2}{3}$	$\frac{4}{6}$	$\frac{6}{9}$	$\frac{8}{12}$	$\frac{10}{15}$	$\frac{12}{18}$	$\frac{16}{24}$	$\frac{20}{30}$
$\frac{1}{4}$	$\frac{2}{8}$	$\frac{3}{12}$	$\frac{4}{16}$	$\frac{5}{20}$	$\frac{6}{24}$	$\frac{8}{32}$	$\frac{10}{40}$
$\frac{3}{4}$	$\frac{6}{8}$	$\frac{9}{12}$	$\frac{12}{16}$	$\frac{15}{20}$	$\frac{18}{24}$	$\frac{24}{32}$	$\frac{30}{40}$

This set tests pupils' understanding of equivalent fractions. Other packs are possible for other topics.

Variation There are lots of possible variations. In one version, which can have more than two players, the players are dealt cards with a number from 1 to 9 on them. Each player turns over a card. If the sum of **two successive** cards is 5, 10 or 15, the first player to call 'SNAP' and call out the two numbers is the winner and keeps those cards. The winner is the one with the most cards at the end of a certain time, or when the other players run out of cards.

5 Pairs For 2 or more players. A pack of special cards is required (the same cards that were used for Snap).
1 The cards are spread face down on a flat surface, with no two cards touching or overlapping each other.
2 The first player turns over any two cards. If the answers on the two cards are the same then he keeps the two cards and has another go. If the cards have different answers, the two cards are turned back over and left in place.
3 The next player now turns over two cards and if they have the same answer, she keeps them, and so on.
4 When all the cards have been paired, the winner is the player with the most cards.

> ***Activity***
>
> Try out these games. Experiment with different topics on the cards and with changes to the rules. Discuss which games you like. Try them out with your pupils.

7 Books with ideas for mathematical games

Many books and magazines include good maths games. These are a few suggestions.

Garrard, W., *Starting a Mathematical Games Library*, Crosslinks Publications

Kirby, D., *Go Further with Games*, Collins Educational, London, 1992

Palmer, R., *Springboard Mathematics,* Nelson, Australia

Rice, T., *Mathematical Games and Puzzles,* Batsford, London, 1973

A14 Materials for teaching mathematics

1 Introduction

Maths teaching requires a lot of materials. Apart from the textbook, these materials need not cost a lot of money. Most of the materials needed to teach maths can be found easily, and freely in the local environment. This unit looks at the materials needed to teach maths, and how to use them.

2 Textbooks

The textbook is there to help the teacher. The teacher should not feel controlled by the textbook. The textbook is a source for the teacher to use. It can give him or her:

- **examples, problems and exercises for classwork and homework**
  Examples in a textbook are usually **graded.** The examples at the start of an exercise should be easier than later examples.
- **methods of teaching**
  Should the teacher use the same approach as the textbook? The teacher has to decide if he/she likes the approach.
- **explanations**
  The teacher has to decide whether the explanation in the book is the best one. There may be an easier alternative.
- **all the main ideas of the course**
  The book should contain the important definitions, procedures, properties, notations, and conventions required for the course.
- **an ordering of the topic**
  The book will teach the topic in a certain order. Is the order good? It may not be suitable for everyone.
- **teaching material when the teacher is not with the pupils**
  For example, when doing homework. A pupil who was absent or who needs a reminder of what happened in the lesson should be able to read and understand the textbook.
- **revision material**
  The pupils do not need to copy lots of notes into a workbook. The textbook may also contain special revision exercises.

3 The chalkboard

This is an important resource for the teacher. Teachers need to learn to use it well. A useful board for the maths lesson is a board with a square grid on it, for use in statistics, coordinate geometry and number work. The grid can be made with a sharp instrument or by using paint.

A collection of **templates** is also useful. These are pieces of card that are in the shape of squares, rectangles, etc. that the teacher can use to draw around and make accurate shapes quickly when teaching.

Activity

Questions to consider when using a textbook: Discuss these questions with your colleagues with reference to the textbooks that you use with your pupils. If the material in the book is not sufficient, prepare material to fill the gap.

- How well does the textbook fit the syllabus?
- What extra material is needed to cover the syllabus?
- Does the textbook contain enough material for the less-able and the very-able?
- Is there enough enrichment material for all levels?
- Are more review and revision exercises needed?
- Are there enough exercises?
- Are the explanations in the textbook clear and easy for pupils to understand?

Activity

Questions to ask when buying a textbook: If you have to choose a textbook there are several questions that can be asked.

- How expensive is the book?
- How are topics organised? Are they taught in one section or broken into parts spread through the year?
- Is there a workbook? Is the book attractive?
- Does the book show maths relevant to your culture and society?
- Is it easy for your pupils to read, with simple language and short sentences?

Activity

Make model cubes, cuboids, tetrahedrons, and so on, for use in school. Make them strong and able to last a long time.

4 Models

A school needs to have a collection of cylinders (empty tins that have been washed), circles, cuboids (boxes), and so on. Keep them together in a storeroom. Be on the look out for containers with interesting shapes such as pyramids and triangular prisms.

5 Card and paper

Card In many topics and games, card is very useful. Keep all the card you can find — the backs of writing pads, old boxes, and so on. There is a lot to be found and it always can be put to use.

Tracing paper/thin paper Tracing paper is very useful for making tracings of shapes that you want to copy or rotate. Thin paper that can be seen through is easier to obtain and does the job just as well. Collect it and store it for later use.

Left over paper Paper that is unused on one side is very useful for all sorts of activities and topics, for example, fractions and symmetry. Many quadrilaterals can be made from a sheet of rectangular paper.

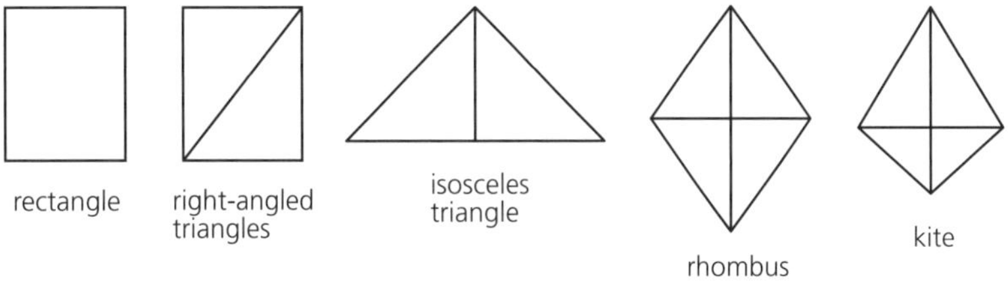

All the properties of plane figures can be explored by using pieces of old paper. A4 size paper is very useful. Keep all sheets that only have writing on one side. Also keep old large envelopes, which can be used for storage, etc.

6 Maths library

Collect as many books about maths as you can. Good sources of books are the Ministry of Education, foreign embassies and high commissions, local charities, retiring teachers, etc. The more books about maths you have, the better!

7 Storing materials

It is important to store materials well, otherwise you will find that you have to keep making new equipment. Keep materials you and the pupils have made in labelled boxes and bags. Old envelopes are very useful for keeping games in. Containers are also needed to store small things in such as stones and sticks. Keep card and paper in a box so that it is stored flat and easy to use.

Try to get a room or cupboard you can use to store this free equipment.

A15 Calculators

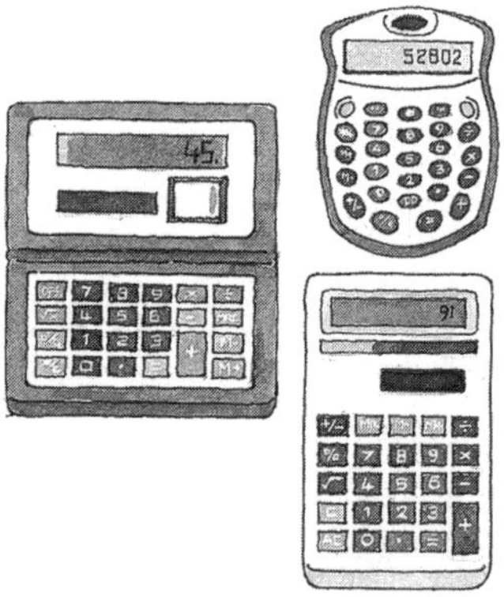

Some inexpensive calculators use solar power and do not need batteries

1 Introduction

The pocket calculator is now used everywhere in the world, but in many countries it is not used in the Primary school classroom. What is the place of the calculator in the Primary school classroom today? Should the calculator be allowed in the Primary maths classroom? The purpose of this unit is to discuss this issue.

2 Helping the less-able

Some writers suggest that maths is difficult for many less-able children because what is learnt in maths requires knowledge of what was learnt earlier. Maths is rather like a large house, where the upper floors are on top of the lower floors, and these are on top of the foundations. If the foundations are weak, the house falls down.

For example, when a pupil starts to learn long multiplication he/she has to have already mastered multiplication tables. However, less-able pupils find it difficult to learn multiplication tables, and so they cannot advance with success to the next stage. Researchers have found that if these pupils are allowed to use calculators for these basic number facts, then they can advance. Of course, we want them to eventually manage without calculators, and to know the tables. They will be motivated by their success, and the repetition of the basic method of doing long multiplication will enable them to learn more quickly than they would have done if they were continually failing.

3 Making it possible to use realistic numbers

The trouble with practical work is that it often leads to 'difficult' numbers. For example, if pupils measure the circumference and diameter of a tin, and divide C by d to find an approximate value for π, it is much better than just telling them a value. However, the pupils are often left with calculations like $21.2 \div 6.8$ to work out. These are difficult and reduce the attractiveness of practical work. The purpose of the practical exercise was not to practise long division, but to find a relationship between C and d — a calculator will reduce the amount of tedious calculating that needs to be done. The availability of a calculator makes it possible for the teacher to use realistic, practical examples without worrying about them giving rise to complicated computations that pupils spend a lot of time working out. Here are some examples of calculations easily done with a calculator.

- How many seconds are there in a day?
- How many months, days, hours, minutes, seconds have the pupils been alive?

(Ignoring leap years, the calculation for seconds for a nine-year old would be $9 \times 365 \times 24 \times 60 \times 60$.)

- How many times do we breathe in a day, a year, a lifetime?
  (The pupils need to count the number of breaths they take during one or two minutes to make an estimate.)
- How many blades of grass are there in a football field?
  (The pupils need to sample a small area first, and know about area as well.)
- How much do all the children in the school weigh? How many elephants weigh as much? How many buses?

Activity

Think up with some interesting questions that could be tackled using a calculator.

All sorts of *how many?* and *how much?* questions can be asked. Ask the pupils to estimate the answers before they work them out. Let the pupils come up with their own questions, and then discuss what is needed before the answer can be calculated.

4 An aid to learning about patterns and numbers

The calculator can be used to enable children to investigate number patterns and develop number fluency. Here are some examples:

(i) The calculator can be effectively used with a hundred board.
Ask pupils to press 5, +, 5, +, 5 … repeatedly, and shade the numbers that result.
What is the pattern? Two straight lines of numbers ending in 5 or 0.
(That is the test for divisibility by 5.)

(ii) Repeat with other numbers, such as 9.
What is the pattern? Diagonal lines across the board.
Notice the sums of the digits of the numbers.
(That is the test for divisibility by 9.)

(iii) Repeat for 2.
What is the pattern? Five straight lines of numbers ending in 2, 4, 6, 8, or 0.
(That is the test for divisibility by 2.)

(iv) A referee chooses a two-digit number as a 'target,' and four single digits as 'ammunition.' The players try to hit the target using only the four digits and the four operations (+, −, ×, ÷).
Score
10 points for hitting the target,
5 points for scoring one more or one less,
2 points for scoring two more or two less.
For example: target 45 and the ammunition 3, 5, 7, 8.
$7 \times 5 + 8 + 3 = 46$, scores 5

(v) You are only allowed to press the 3 and 4 buttons and the four operation buttons. Use them to make each of the following numbers on your calculator: 9, 10, 11, 16.

For example: $4 + 4 + 3 = 11$ $\quad$ $3 + 3 \times 4 + 4 - 3 = 16$

This exercise examines pupils understanding of basic principles. If they get the answer wrong, it is easy for them to try again and learn from their mistakes. Different calculators use different conventions for arithmetic. This exercise helps the pupils understand how their calculator works. The result for the second example may be different on some calculators.

5 Checking answers

The calculator can be used to check that arithmetic has been carried out correctly.

6 Teaching children to use calculators properly

All calculators come with an instruction book, and children can learn how to use calculators well. The calculator is not meant to make them lazy — it is meant to help them calculate. By taking the tedium out of long calculations, they can concentrate on the mathematics of what they are doing. They should not need a calculator for basic arithmetic, (although some pupils will find it helpful) but it helps them develop a better understanding of numbers.

A16 Assessing mathematics in the Primary school

1 Introduction

Effective assessment provides information and guidance for:

- **the learner**
  (i) provides a statement of success in meeting learning objectives
  (ii) confirms his/her strengths and diagnoses his/her weaknesses
  (iii) can provide an incentive to continue learning
- **the teacher**
  (i) provides a statement of success in meeting learning objectives
  (ii) diagnoses strengths and weaknesses in teaching
  (iii) provides feedback which can be used to regulate the pace of teaching, or to reconsider the methods used
  (iv) indicates the extent of remedial work required
  (v) puts pupils in order, which can be compared with earlier/expected orders
- **the Principal or Headteacher**
  (i) measures effectiveness of a teacher's work. (Interpretation has to be done with care; it is important to take into account the ability and background of the class. It is important to recognise how much a teacher has improved the standard of a class **value added** rather than take a crude measure of a class' ability.)
  (ii) provides a means of monitoring individual classes and departments
- **parents**
  (i) gives a measure of their child's progress and development
  (ii) enables them to assess the quality of the teacher and the school
- **the Ministry of Education**
  (i) measures effectiveness of teachers' work
  (ii) provides feedback on the curriculum
  (iii) enables comparison of standards (boy/girl, rural/urban, etc.)
  (iv) helps in awarding bursaries
  (v) facilitates awarding of places in secondary and tertiary education.

> ***Activity***
>
> Discuss whether it is fair to judge a teacher's performance by the success of his/her pupils in exams and tests.

2 Formative and summative evaluation

Pupils are regularly assessed through classwork, homework, tests, school exams and public exams. These assessments can be divided into two categories of evaluation, named by Bloom **formative** and **summative** evaluation.

Formative evaluation

Oral and written classwork and homework are used by the teacher to find out whether pupils understand what they are being taught. These are examples of **diagnostic tests.** Teachers diagnose what is wrong with their pupils' mathematics and they form an opinion and decide on a course of action. This method is used intuitively throughout the learning process. Its main role is to discover successes and failures in learning and to provide feedback that can be used to improve the course.

Summative evaluation

End-of-year or public examinations sum up how pupils have done overall in a course. Summative evaluation usually occurs at the end of a course or even a topic, often to provide a score for administrative or statistical purposes.

Summative evaluation influences teachers' methods. They often cannot afford to spend time on interesting side issues, but must concentrate on teaching for the examination.

Note that many tests are poor indicators of the level of a pupil's understanding, or of his/her mathematical potential or reasoning power.

Formative evaluation	Summative evaluation
• Has the pupil learned enough to begin the next stage? • How well is the group as a whole acquiring the specific skills and knowledge taught? • What remedial work is needed, and for whom?	• What is the pupil's position in the rank order of the group? • How well has the group as a whole acquired the broad range of skills and knowledge taught? • How well does the group conform to agreed standards?
Content • Usually concerns a narrow and specific area.	**Content** • Tests usually cover a wide cross-section of topics.
Mark discrimination • Most pupils should score high marks if the test is appropriate and learning effective. • This is important for class motivation.	**Mark discrimination** • Pupils who have learnt well should do well. • Weaker pupils should get low marks.

3 Formative evaluation — homework marking

It is important to set *and mark* homework, because regular marking and the recording of marks shows the pupils that the teacher is interested in what they are doing. The marks gained and recorded give a good record of how well pupils have learned.

When setting homework do not make the work too hard or too easy. Set homework carefully with a range of difficulty that it is easy for all pupils to get about 50% but quite hard to get 100%. The range of difficulty means that all pupils have something to challenge them. If all questions are too easy, then the best pupils will become bored. If all questions are too hard, many pupils will be discouraged.

What can be done to ensure that pupils do their homework? Many pupils come from homes where it is difficult to work in the evening. Some pupils go home and do not want to study. Here are some suggestions:

- Collect homework in and mark it. If the homework is not marked, and merely gone over in class, there is little incentive for the pupils to do it.
- Keep a mark book so that there is a record of who has done the work and how well they have done. Pupils do not like having zeros against their names.
- Follow the advice given earlier of not making it too difficult.
- Give a variety of activities. A homework could be to measure something at home, make a model, etc., it does not always have to be an exercise of problems.
- Do not make the homework too long. Do not set it every day. Two or three days a week is enough.

Activity

In some countries it is illegal to give pupils under the age of nine homework to do. They are recognised as children learning to socialise at this stage. At what age should homework start? Discuss the measures that you can take to make sure that your pupils do their homework.

If the pupils have only one exercise book for homework and classwork, it is difficult to collect work in and mark it. If they have a special book for homework, it can be collected in the following day and they still have a book for working in during the lesson.

4 Changes in assessment procedures

Traditionally it has been the practice to test pupils in formal written exams. Many people question whether a written test taken in stressful conditions is the most effective way of finding out how much pupils have learnt. Written exams emphasise the solution of routine problems. They place little emphasis on problem-solving skills, and encourage teachers to teach maths as a collection of techniques designed to answer exam questions, without thought or enjoyment on the part of the pupils.

One central objective of many syllabuses is the ability to solve realistic problems. The routine problems of exams do not encourage this. Project work can develop this skill. Teachers should ask their pupils to undertake projects in the maths class. Section A, Units A11, A13 and A15 look at some examples of projects that can be used in Primary school and that require few resources.

A17 Testing mathematics

1 Introduction

This unit looks at the writing of tests. Tests are an important part of maths teaching. Tests should be well prepared, so that the pupils are able to show what they know as effectively as possible.

2 Setting a test – things to keep in mind

Tests should have a **balance** of items. There should not be too many questions on one topic. There should be a **range** of difficulty. One way to create a balance of questions is to use a **grid** to categorise the questions. Here is a grid for a test on fractions:

Fractions	Calculation	Problem	Non-routine
Equivalence	1	–	–
Addition	2, 3	10	–
Subtraction	4, 5	11	–
Multiplication	6	12, 13	–
Division	7, 8	–	15
Mixed	9	14	16

Other things to consider are:

1 Are the questions the kind that the pupils are used to?
2 Is the wording of the questions ambiguous?
3 Use clear simple English — short sentences, simple constructions.
4 Know how much time the pupils will have, and write the test to fit that time.
5 Know which topics you wish to test, and make sure the questions cover those areas.
6 Ask questions about important things, not about unimportant things.
7 Start the test with easier items to warm the pupils up.
8 Check the test by doing it yourself before you give it to the pupils. This will often reveal errors. (Better still, ask a colleague to do the test before you use it.) It is very difficult to produce an error-free maths test. In maths, the precise use of symbols and words is so important.
9 Do not make the test too hard. Try to have everyone scoring 30% or more. Weaker pupils may feel that they are not making any progress if they get a very low mark.
10 On the other hand, include some harder questions at the end of the test to challenge the brighter pupils. This will prevent them being bored, and over-confident.
11 On the test paper, state the number of marks given for each answer, especially if you have long questions in the test, so pupils can judge how long to spend on each item.

Activity

Write a test for a topic of your choice. Use a grid like the one above to make sure that you have included a balance of items and a range of difficulties.

3 Cognitive levels of test items

Test items can be classified into four different levels.

Level 1 Knowledge recall items

These require the use of memory only. The pupil is asked to recall something in the form in which he/she originally learned it.

For example: $2 + 3 = \ldots?$ What is the angle sum of a triangle?

These questions are easy to construct and mark. Usually they only test rote learning and memory, and are not often used in maths exams.

Level 2 Routine calculation items

These require pupils to display the ability to follow or use a technique which is either given or they have memorised, such as being able to manipulate expressions and to use mathematical instruments.

For example: $327 \times 32 = \ldots?$ $846 \div 27 = \ldots?$

These questions are easy to construct and mark.

Level 3 Routine problem items

These are at the lowest level at which relational understanding is required. The pupil is required to show understanding of a concept by using it directly in a familiar situation.

For example: Find the selling price of a dress bought for $20 and sold at a profit of 50%.

These questions are easy to write. Marking has to allow part marks for answers that are only partially correct (through arithmetic errors). These questions can test routine and non-routine *situations*.

Level 4 Non-routine problem items

These should provide a challenge with questions that are unfamiliar to the pupil. They may require the pupil to put two or more concepts together to arrive at a solution. These questions can test routine and non-routine *situations*.

For example: In my drawer there are 10 black socks, 8 red socks and 6 white socks. What is the smallest number of socks that I need to take out if I want to be sure that I have:
(a) two socks of the same colour, (b) two red socks.

4 Types of questions

Questions can be posed in a variety of ways. Varying the format of tests helps to keep them interesting.

Essay questions are unusual in maths. They are easy to set but hard to mark.

Matching item questions can be set at any level.

For example: Match these two lists.

whole number	0.3
fraction	5
decimal	$\frac{1}{2}$

Multiple choice questions are difficult to set but easy to mark. When there is a large number of pupils taking a test, a multiple choice test saves a lot of time on the marking. Multiple choice questions have a stem question (underlined in the example below) and several (usually 3 or 4) possible answers to choose from. A multiple choice question will usually include *distracters*, that is, answers that the pupils may arrive at if they make one of the common errors.

For example: $\underline{0.6 \div 0.01 = ?}$ **A** 0.006 **B** 6 **C** 60 **D** 600

C is correct. **A** is for pupils who assume they should 'count the total number of decimal places on the question'.

A good multiple choice question has:

- a clear, concise statement of the central problem
- plausible distracters.

Always ensure that one of the answers is correct. This problem has no correct answer listed: If $-4x + 5 = 3$, $x = ?$ **A** –4 **B** 5 **C** 3 **D** 2

Avoid two correct answers, as in this example:

Describe a shape that measures 2 cm by 2 cm.

A circle **B** square **C** triangle **D** rectangle

Avoid trick distracters. For example:

The volume of a box 4 cm by 3 cm by 2 cm = ?

A 24 cm^2 **B** 24 cm^3

Think carefully about your choice of good distracters. For example, a good distracter for $\frac{1}{2} + \frac{4}{7}$ would be $\frac{5}{9}$.

The distracters chosen must make the choice of answer challenging. For example in:

$269 + 7 - 234 = ?$ **A** 0.4919 **B** 9.081 **C** 42 **D** 100.395

only one answer, C, is remotely possible, the pupils could choose this by guesswork.

Similarly, $25 \times 12.3 = ?$

A 4 **B** 6.7 **C** 12.3 **D** 307.5 **E** 10 000

Avoid unnecessarily complicated problems. For example, $7\frac{1}{8} \div 4\frac{5}{9}$ is not suitable for pupils just introduced to division of fractions.

Avoid emotive ideas and issues — wars, orphans, death, etc.

Avoid the letters A, B, C, D and E in the stem. They can be confused with the letters used in the answers.

Have a resource of books, old questions, etc. for ideas. Look out for the kinds of errors that pupils may make, and use them as distracters.

Activity

Make up 4 multiple choice answers (one must be correct) for each of these questions.

1. $\frac{3}{7} \div \frac{4}{5}$
2. Add 4 to 98.
3. What is 5(8 – 3)?
4. Work out $(^{-}4) - (^{-}2)$.
5. $2\frac{1}{4} \times 3\frac{2}{3}$
6. $\frac{1}{4} \div 2$
7. 6 books cost \$4. How much do 8 books cost?

A18 Preparing pupils for examinations

1 Introduction

All teachers want their pupils to do as well as possible in exams. The purpose of this unit is to look at how maths teachers can help their pupils prepare for maths exams in the most efficient way. A lot of pupils do not prepare for exams properly. Consequently, they do not do themselves justice, and get a lower grade than they had expected. Teachers should offer their pupils advice so that they can do their best in exams.

2 Revising

Merely reading the exercise book and the textbook is not a very good way to revise. Many pupils do this and learn very little because their revision is passive and not active. Quite often, they fall asleep doing it! Try to make your pupils realise that it is very important that they practise as many exercises as possible when revising mathematics. There are two reasons for this:

- Doing the exercises will tell the pupils whether they have understood the topic fully. If they have not, then they should go back and read through the topic again.
- Mathematics examinations consist of problems. Doing exercises prepares pupils for the exam questions they are going to meet.

Here is a suggestion for a **revision strategy** that pupils should follow:

- Look at the syllabus topic by topic.
- Read through a section of the textbook or exercise book for that topic. Try to identify the important information about the topic. Then close the book and try to write down the important information. Open the book again and compare what you have written with what is in the book.
- If you are mostly correct, work through the examples. Try to do each example before reading the solution in the book. If you cannot do it, read through the solution in the book. Then try it again without looking in the book.
- Do some exercises for the topic.
- Do some questions on the topic from past papers. However, do not do them until you have revised the topic. They may seem rather hard and put you off.
- Look at past papers to make sure that you are familiar with the way the exam is organised. Know the format for the exam. Know the instructions.

3 The day before the examination

The pupils need to be as well as possible. It is important that they are feeling their best when they go into the examination. Tell them to make sure that they get plenty of sleep the night before the exam.

They should not attempt to do a lot of work at home on the night before the exam. (They should have revised thoroughly in plenty of time.) They want to go into the exam hall feeling awake and at their best, so that they can do their best.

4 Doing the examination

At the start of the exam, it is a good idea to write down all the formulae that are not stated on the exam paper that you may need in the exam. It is very easy to forget these when under the pressure of answering a question.

Read through the paper carefully. Underline the important words in each question.

If you cannot do a question, you should not panic. Go on to the next question, and come back to it when they have finished the other questions. Examinations can be very frightening. It is easy to panic or to get stuck on questions. The best approach is to stay calm and attempt the questions you can, then go back and try the ones you have missed out or partly done.

If you cannot finish a question, do not cross out the work you have done. It may earn some marks. In any question, do what you can and you may pick up some marks.

Do all working in the space provided on the answer paper next to the question. Sometimes marks are awarded for being partially correct. If all the working is close to the question, the marker will be able to see it and may give some marks for a part solution, or for using the correct method.

Set out work neatly. If the work is messy, it is easy for you to misread your own working and get the answer wrong. It is also hard for the examiner to understand your working and give marks for the use of a correct method. The examiner cannot give marks if he cannot read the writing.

Read the question carefully. Ask yourself these questions: '*What information is in the question?*' '*What am I being asked to work out?*' '*What formula do I need?*' Write the formula down as you have remembered it, and then substitute numbers in it.

If there is a **question about angles** on a figure, and you cannot immediately see what the answer is, write all the angles you can on the figure. Often the answer to the question will then be easier to see.

Make a **rough estimate of the answer** to calculation questions. This will help you check whether you have made a careless mistake when calculating.

Make sure you give an answer in the correct **units.**

Make sure that answers are given to the required amount of **accuracy** — the correct number of decimal places or significant figures. When solving a problem, do not approximate too early as this can change your answer too much.

Time is important in exams. Be aware how long the exam is. Never spend too long on a single question. There may be easier questions later in the paper. You can pick up many marks from them. If you spent too much time on the earlier questions, you will not have time to get to them.

You may find that you are **not able to finish** the whole paper. If this is the case, you need to try to get as many marks as possible. So, when there are about 20 minutes left to the end of the exam, look through the questions that remain to be done and pick out the easiest parts of questions and do them first.

Activity

Discuss the suggestions above for improving exam performance with your colleagues. Are there any additions or amendments that you would like to make?

There are no extra marks for **finishing early**. If you finish early, use the time to go through your paper and check your answers. Check whether you have made any careless mistakes.

The number of marks can differ from question to question. If a question is only worth a few marks, do not spend very long on it. Spend longer on questions worth more marks.

Section B Mathematics topics — number and algebra

Topic 1 Pre-number work

1 Introduction

Often, children learn about numbers at home, before they come to school. Some will have learnt to use 'one' and 'two'. Relatives will have taught them how to count. Many pupils will know their ages. In many cases, however, the children will only know how to recite or sing 'One, two, three, four...' without being able to count things meaningfully. This first topic looks at activities designed to help children to be ready to count.

2 Preparing children to learn about numbers

Give the pupils collections of objects to sort. For example:

- pictures cut out from old magazines and newspapers to sort into cars, animals, houses, men, women, etc.
- shapes to sort into round, square, etc.
- shapes to sort by their colour
- bottle tops to sort by their brand names (or designs).

Activity

- Collect used drinks cartons. Wash them and make them into sorting trays.
- Collect shapes of different colours.
- Collect various soft drink bottle tops.
- Collect storage containers for these items.
- Let the pupils practise sorting them.

Use sorting trays made from empty drink cartons with the tops cut off.

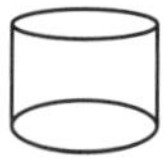

The pupils can:

- **describe** the sorted objects, using words like *round*, *corners*, *red*, etc.
- **compare** the sorted objects, using words such as *more*, *less*, *as many as*, *larger*, etc.
- **match** sets with the same number in them.

Young children often confuse the number in a set with the way it is arranged.
For example:

Children often think that Set *B* has more counters in it because it is longer.

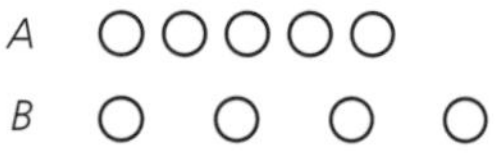

This can be avoided by using squared paper.

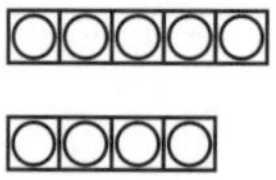

An abacus can be used for comparing.

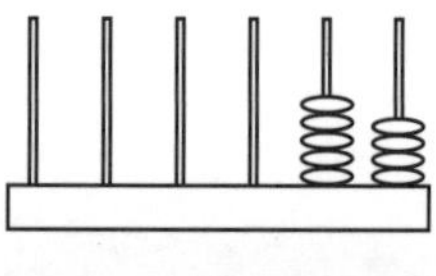

The counters on the abacus can be used to *represent* (an important idea, but not a word to tell to the children at this stage) the number of objects in a set. One bead is used for each object being counted. The set with more members has more beads.

Activity

Find pictures in magazines and newspapers suitable for pupils to sort. You will need a lot for a whole class. If a class of 40 pupils work in pairs, you need 20 sets of about 15 pictures, that is 300 pictures! Keep them in a cardboard box for future use.

Topic 2 Introducing numbers

1 Introduction

Numbers are about sets not individual objects. When we say that there are five toys, we are talking about the set of toys, not the individual toys. If there were more toys, the number would change but the toys would remain the same.

Cardinals and ordinals

Cardinal numbers (1, 2, 3, 4...) specify the size of a set. The last number that you say when counting them is the number of things in the set.
Ordinal numbers (1st, 2nd, 3rd, 4th...) specify position or relative size.

> ***Activity***
>
> Make a collection of rhymes that can be used to introduce small numbers.
>
> Ask your parents, your colleagues, and your friends for examples.
>
> Draw simple pictures to illustrate the numbers 1 to 10.

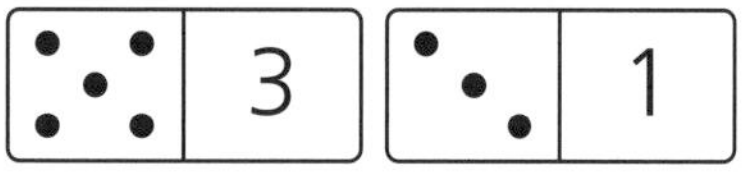

2 Learning to count — small numbers

The number 1	Children will have learnt the meaning of 'one banana', 'one cup', and so on, before coming to school.
The number 2	Many will have learnt about their two legs, two eyes, and so on. Important words are *couple, pair, both.*
The number 3	Show pupils examples of 3 in pictures and with three things.
The number 4	Show pupils examples of 4 in pictures (4-legged animals) and with four things.
The numbers 5 to 10	Pupils can count 5 fingers, 6 objects, 10 toes, and so on.

Number rhymes help pupils learn the numbers.

Dominoes and dice are useful tools to reinforce the association between numerals, amounts, and the names for numbers up to six. You could make a set of dominoes from thick card that combine dots and numerals.

3 Learning to count larger numbers

The numbers from 11 to 99 are learnt as names at this stage. 56 is just the name for a number — thinking of the number as 5 tens and 6 ones comes later on. '75' is a way of writing a number name. 40 is not understood as being 4 tens but as being 'forty'.

Associate numbers with things familiar to the pupils — 7 players in a netball team, 11 players in a football team, dates of birth, and so on.

Children often learn to recite before they can count. Ask the pupils to recite numbers in order — 1 to 12, 13 to 19, 20 to 99. Some will get stuck at 29, 39, etc. until they realise the pattern of counting. Pupils need to match the words to the number of objects. It helps if they *touch* the objects when counting.

Pupils can practise counting as the teacher goes round the class with each pupil standing up in turn and saying the next number.

4 The number line

The number line is very useful in learning number skills. There are two ways of showing the numbers:

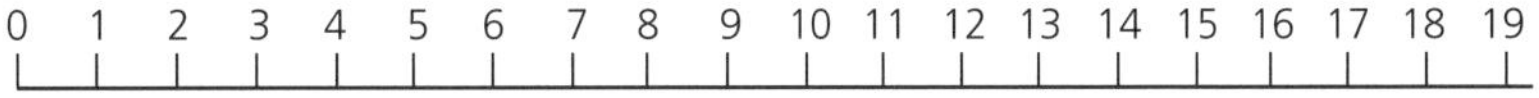

0	1	2	3	4	5	6	7	8	9	10	11	12	13	14	15	16	17	18

In the first, the numbers are marked by a point on the line. In the second, the numbers are shown in a space on the line The number line can help pupils to see that the distance from 0 to 10 is same as the distance from 30 to 40. It also helps to show that 38 = 30 + 8 and that 46 is between 40 and 50.

5 Comparing numbers

Which is larger, 18 or 21? Which is smaller, 14 or 16? Questions like these are answered at this stage by counting. The number line can be very helpful, too.

6 Estimating numbers

Pupils can play a game to estimate the number in a set. Give them a pile of things. Ask them to estimate how many there are, then ask them to count the objects. Estimating is an important skill in much of the later maths they do, so it is good to start estimating early, in a simple way.

7 Numbers important in the child's life

Some numbers are important in a child's life. Use them in class. Try to persuade parents to talk about these numbers at home. Some of the numbers that can be discussed with pupils are:

- dates, ages, times of day
- birthdays of family members, the number of days to a child's birthday
- distances on signposts
- counting the number of pupils on the class register.

More numbers can be introduced by:

- playing games using dice, recording scores in games, etc.
- displaying scales, height charts, etc. in the classroom.

Activity

1 Collect information about the number of members of different sports teams. Discuss reasons for the different number of players in different sports teams.

2 Make charts for display showing 3 birds, 4 oranges, and so on.

3 Make a set of dominoes like the ones shown on page 60.

8 Learning to write numbers

Children need to learn to write numerals neatly. With the base ten system, there are only ten symbols to learn to write. In the diagram below the symbols are divided into those numerals that require only one stroke and those that require two strokes. The dot indicates where the stroke should start from in each case. The numbers 4 and 5 require two strokes of the pen.

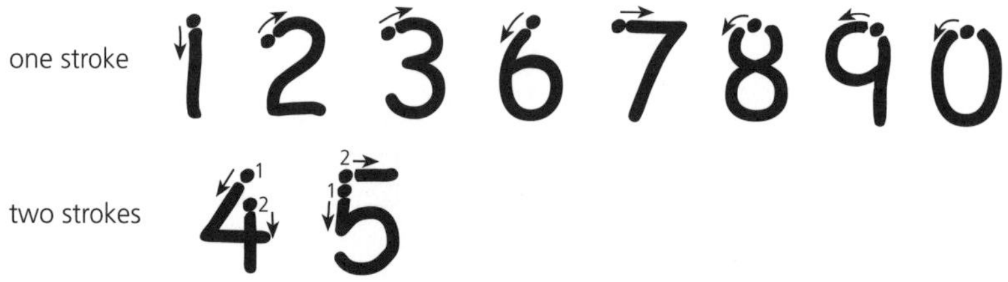

Topic 3 Place value

1 Introduction

The **Hindu-Arabic numerals**, which were developed in India and spread by the Arabs, have been adopted all over the world. If you look at a Japanese mathematics textbook all the symbols are the same as ours and we can understand the mathematics even if we do not know a word of Japanese. The popularity of the Hindu-Arabic numeral system arises from its power and simplicity. Only a small number of symbols are needed to express any number, however large or small it is. These symbols are, of course, 0, 1, 2, 3, 4, 5, 6, 7, 8, 9 and the decimal point. The **value** of a symbol depends on its position in the number. Thus the 5 in 25 436 stands for five thousands, whereas the 5 in 0.005 stands for five thousandths.

2 Teaching place value

If your pupils are going to learn how to carry out the four operations on numbers they must understand place value thoroughly. For this, they need to have a lot of practical experience. Several different teaching aids can help to teach place value and the four operations.

- **Bundles** Short sticks are used for ones, and bundles of 10 sticks are tens. Ten bundles of 10 sticks are hundreds, and so on. The number shown here is 23:

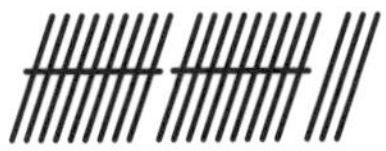

- **Squared paper** This equipment is based on Dienes' blocks which are named after Zoltan Dienes.
  A small paper square is a **one**:
  Ten small squares in a line are a **long** or a **ten**:
  Ten longs make a **flat** (or **hundred**):
  You can make a **box** (one **thousand**) whose faces are all flats but this is not necessary to teach the principles of place value and operations with number.
  The diagram below shows the number 243 using flats, longs and squares (hundreds, tens and ones).

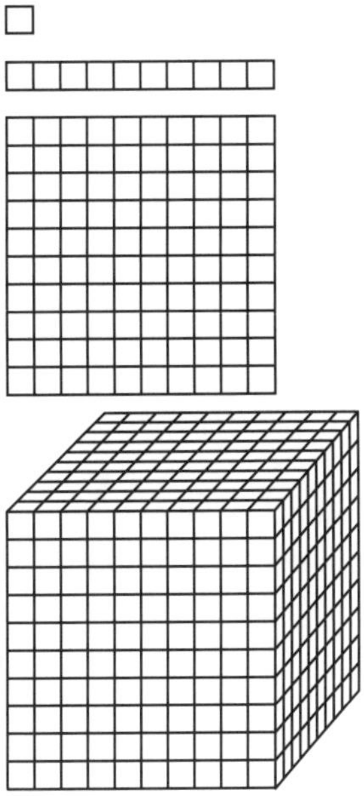

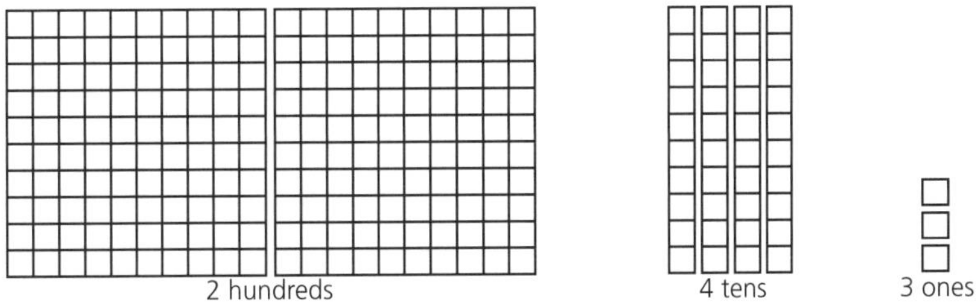

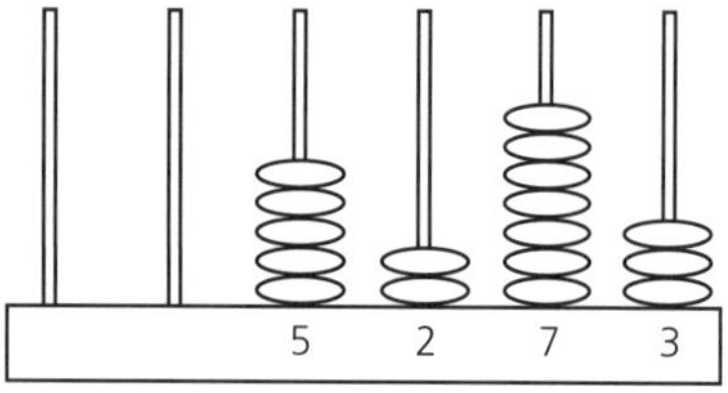

- **Abacuses** Rows or rods of beads (9 to a row) are on wires or sticks. Each bead on a rod is worth 10 of the beads on the rod to the right of it.
  This abacus shows the number 5273.

Note that there is a major difference between the abacus, and the bundles and squared paper. A bundle of 10 **is** 10 times the size of a one, a long **is** 10 times the size of a square, a flat **is** 100 times the size of a square. In these cases, the size is *proportional* to the number it represents. They represent the numbers 'structurally'. however, on the abacus a ten-bead is not the same size as 10 one-beads. The one-bead and the ten-beads are in fact the same size. The *size* of the bead does not represent the number, but its *position* tells us the value of the number.

Activity

1 Collect a couple of hundred short sticks of about the same length and make about 100 of them into **bundles of ten** and keep the rest as **ones**.

2 Use squared paper and scissors to make the same number of **squares**, **longs** (from ten small squares in a line) and **flats**.

3 Make an **abacus** with three rods (for ones, tens and hundreds). You will need 27 small beads or rings. The height of each rod should be exactly equal to the height of a pile of 9 beads (so that a rod cannot hold more than 9 beads).

Practise using these teaching aids in the way described in the text. Discuss with your colleagues which method you prefer.

3 Suggested sequence

Teaching aids can be used to teach the following sequence:

1 Start with the numbers 10 to 19. Pupils learn that in the number 17, for example, the 1 means one set of ten and the second digit, 7, is the number of ones (sticks or squares). This can be done using:
   - sticks put in a bundle of 10 and 7 single sticks
   - 1 long and 7 squares
   - an abacus with 1 bead on the second rod and 7 beads on the first rod.

   Show the pupils the number on a number line.

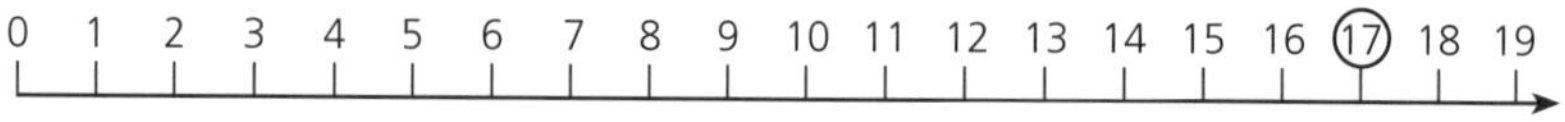

2 Count in tens up to 100. Show this on a number line, jumping in tens, and using longs and bundles.

3 Let the pupils appreciate that the 2 in numbers 20 to 29 means two longs or two bundles and the second digit is the number of squares or ones needed, by making two piles of 10 and adding some extra squares or sticks.

4 Repeat the activity for other two-figure numerals until the child is happy not to do it physically and can look at a number such as 68 and say there are 6 tens and 8 ones.

4 Comparing numbers

Which is larger, 68 or 81? Which is smaller, 34 or 36? Questions like these can be answered by counting. However, that can take a long time. The numbers could be compared on a number line. Pupils can compare the numbers of tens and ones in each number. 81 is larger because it has more tens in it. Pupils should do examples of this, until they can tell you the answer without using the teaching aids.

5 Number games

1 **Estimating numbers game** Pupils can play a game to estimate the number of objects in a set. Give them a pile of things. Ask them to estimate how many there are, then ask them to arrange the things in sets of 10.

2 **Card game** Players need a pack of cards with the digits 0 to 9. Each pupil picks two cards from the pack and makes a two-digit number. The pupil with the largest number wins a point. Repeat 10 times. The player with the most points is the winner.

6 Larger numbers and money

In some countries a single unit of currency does not buy very much. Children have to get used to dealing with hundreds and thousands of kwacha, shillings, and so on, when they go shopping. Teachers in these countries have to introduce larger numbers to pupils quite early, and get them to practise using them.

Activity

Play these games with your friends. (You will need some cards with the digits 0 to 9 on them. Make them out of stiff card so that they can be used many times.)

- Discuss what these games help pupils to understand better.
- Decide whether the rules can be changed to make other games.

Topic 4 Adding numbers

1 Introduction

Addition arises in several ways in number stories. 5 + 3 = 8 can arise from:

- **Combining two sets:** I have 5 toys, you have 3 toys. We have 5 + 3 = 8 toys altogether.
- **Increasing a set:** There were five pupils in the room. Three more arrived. There are now 5 + 3 = 8 pupils present.
- **Comparing sets:** My brother has 5 balloons. I have 3 more. I have 5 + 3 = 8 balloons.

2 A suggested order of teaching

Addition can be introduced in the following ways:

- you have two sets, count the number of objects in each set
- combine the sets
- count number of objects in the new set.

Children usually either **count all** or **count on**.
For example, 4 + 5 = 9. Counting **all**, they put 4 objects and 5 objects together and count 1, 2, 3, 4, 5, 6, 7, 8, and 9.
Once they have got the feel of the numbers, they count **on**. They start from 4 and count on 5, 6, 7, 8, and 9.
First they should develop a feeling for small numbers. Start with + 1, then + 2, and so on.
Later on, move on to larger numbers, such as 8 + 9 = 17.

3 Materials

The basic addition number facts need to be learned by all pupils. They can be established in several ways using the following materials. Let the children use the materials until they no longer need them.

Using small objects, people, fingers, and so on. 4 + 5 = 9, for example, can be established by putting 4 objects (or children) with 5 objects (or children) and counting the total number of objects (or children). Fingers can be used for small sums — especially if the pupils can count on.

Cuisenaire rods are rods that are 1 cm, 2 cm, 3 cm long, up to 10 cm long. Rods of different lengths are different colours. The colours are as follows:

1	**2**	**3**	**4**	**5**	**6**	**7**	**8**	**9**	**10**
white	red	light green	purple	yellow	dark green	black	brown	blue	orange

If the rods are not available, make them out of strips of squared paper and colour them with crayons or shade them.

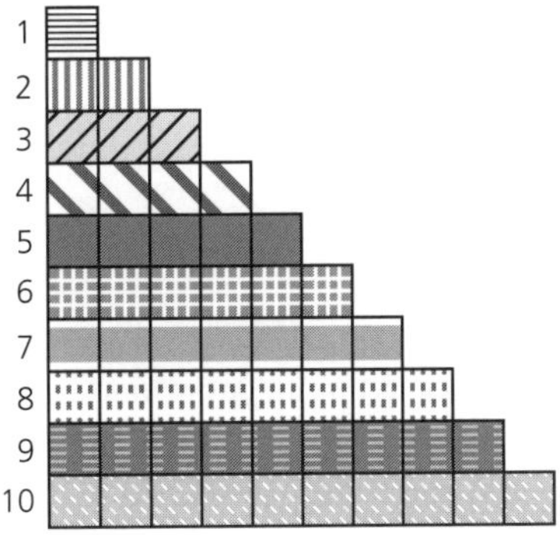

Using the rods, 2 + 4 = 6 can be demonstrated like this:

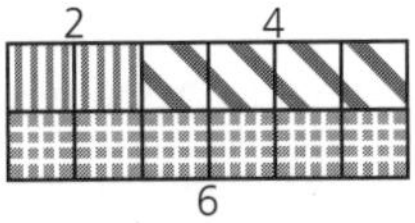

Addition facts can be established using **partitions**.
For example, the rods can be used to show the partitions of 7:

7 = 0 + 7	7 = 4 + 3
7 = 1 + 6	7 = 5 + 2
7 = 2 + 5	7 = 6 + 1
7 = 3 + 4	7 = 7 + 0.

The partitions of 10:

10 = 1 + 9   10 = 2 + 8   10 = 3 + 7, etc.

are important when adding larger numbers later.

Activity

1. Practise using Cuisenaire rods. If you cannot obtain rods, make strips, 1 to 10 small squares in length. Colour or shade each length differently.
2. Make a number road. Practise using it to teach addition facts.

Number road
This differs from a number line. The numbers are in boxes and are not points on a line.

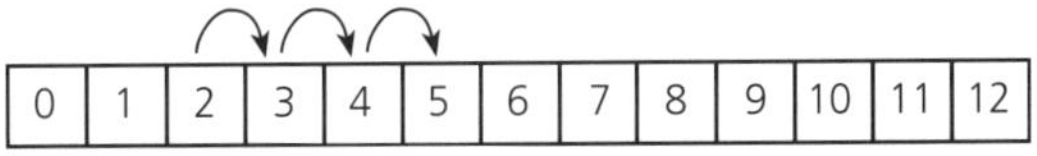

The picture shows 2 + 3 = 5. Pupils could put 12 squares in a line on the ground and number them. 2 + 3 then becomes a jumping game, rather like 'hopscotch.' Go to 2 and then jump three squares to 5.

4 Recording the results

Pupils need to find out the addition facts and *learn* them. They need to record them in a table. The patterns in the table will make it possible for them to complete the table without working them all out. However the pupils have to be able to recall the facts individually.

5 Helping children to learn addition facts

The pupils should learn all the partitions of all the numbers from zero to ten. (0 + 0 = 0, 0 + 1 = 1, 1 + 0 = 1, to 9 + 1 = 10, 10 + 0 = 10) They should also learn all the combinations that make the numbers eleven to twenty by adding single digit numbers, or by adding a single-digit number to 10. (1 + 10 = 11, 2 + 9 = 11, to 9 + 10 = 19, 10 + 10 = 20) This means there are more than 100 addition number facts (including adding 0) to be learnt, and used without referring to equipment. What can be done to make this learning task easier?

- It does not matter which number comes first when adding. For example, 4 + 3 gives the same result as 3 + 4 (= 7). This almost halves the number of facts to remember.
- Adding zero, and adding one (2 + 0 = 2, 3 + 0 = 3, etc., 3 + 1 = 4, 4 + 1 = 5, etc.) are easily learned, which takes care of another 20 or so facts.

- Doubles (3 + 3 = 6, 4 + 4 = 8, etc.) are also easily learned (10 more facts).
- Use the doubles to break down other sums:

  6 + 7 is one more than 6 + 6,
  6 + 6 = 12, so 6 + 7 = 12 + 1 = 13.

Other techniques can be used to help pupils:
- Partitions of 10 can help pupils to remember the larger sums.

  For example, 8 + 7 = 8 + 2 + 5 = 10 + 5 = 15.
  They are important in later work.
- Adding 9 is easier if pupils remember that it is the same as adding 10 and taking 1 away.

  For example, 6 + 10 = 16, so 6 + 9 = 6 + 10 − 1 = 16 − 1 = 15.
- Learning through fun. Practise addition using **games** and **flash cards**.
- Using money is very useful for practical work for mental addition and subtraction.

Group activity

Develop some activities that can be used to help your children develop their memory of addition number facts.

- Are there any songs that can help?
- Can you develop some games?

Topic 5 Addition algorithm

1 Introduction

The calculator has made the ability to work out complicated arithmetical calculations less important. Nevertheless, these skills are still needed and should be taught. It is important that pupils understand why the algorithms work. The emphasis in this work is therefore on pupils acquiring an understanding of algorithms through the use of suitable, simple, practical equipment.

Children start to add small numbers such as $4 + 3$ by counting all, then by counting on, and finally by remembering the result.

$27 + 8$ would first be solved by counting on. It is unlikely we would remember the result is 35, but we do recognise $7 + 8 = 15$. We then combine this with adding 20 to get the final result.

It is not so easy to count on for $47 + 25$. The solution can be found by breaking the problem up: $40 + 20 = 60$, count on 7 and then count on 5 more, is one method.

Simple calculations like these can be done by a variety of techniques which break the problem down into easier, manageable, steps. However, when pupils have to rely on seeing numbers as tens and ones some pupils become confused. The use of practical methods and teaching aids should overcome this.

2 'Carrying' — some general observations

- Writing on squared paper helps children put numbers in the correct columns.
- When there is no carrying it is unimportant which column you add first. However, insist that the pupils add ones first, then tens, then hundreds. This establishes good habits for later.
- Add from the bottom because any number that needs to be carried is less likely to be forgotten. Check the total by adding from the top.
- Write some examples horizontally. This helps children to learn how to put numbers in the correct columns.
- Give some examples as word problems, for example, 'I have 133 sweets; my friend has 22 sweets. How many do we have altogether?' Through such problems, and oral questions, the pupils will meet all the different ways we can talk about adding.

3 Addition of two-digit numbers

The equipment used to introduce place value can be used to help pupils understand addition. It is important to start from simple examples and develop the work in small, manageable steps.

Stage 1 Addition requiring no carrying (answers 99 or less)

- Start with tens and ones in each number, for example 21 + 35. This is 2 bundles and 1 stick, with 3 bundles and 5 sticks, or is 2 longs and 1 square, with 3 longs and 5 squares:

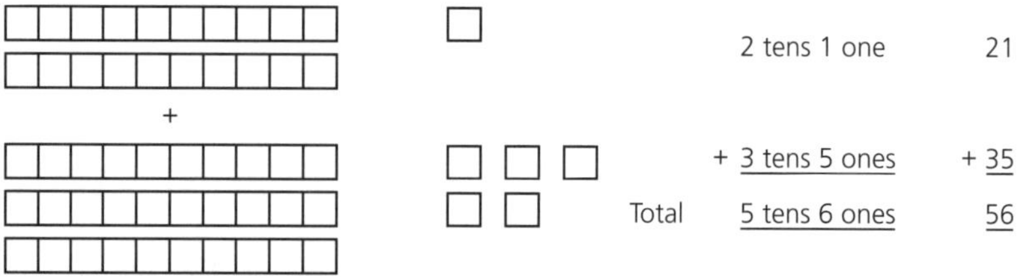

- Later examples should include additions that pupils often find difficult, for example,

  23 + 5 where one number has no tens,

  20 + 31 where one number has zeros in the ones column.

Stage 2 Addition requiring carrying

- Once pupils can work, with or without teaching aids, on examples that do not require carrying, they can start doing examples that do require carrying. Again, start with two figures in each number, for example, 18 + 24.

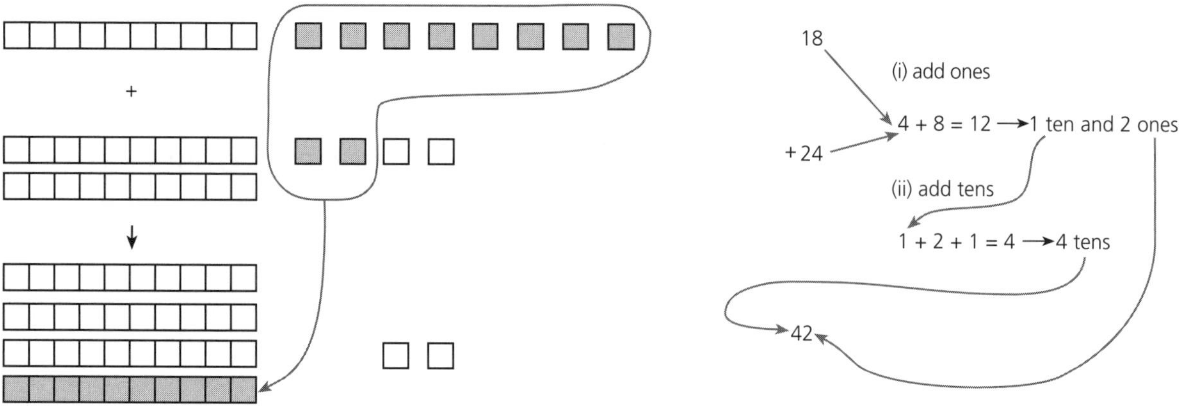

	4 ones + 8 ones is 12 ones.
$\overset{1}{1}8$	10 ones are exchanged for one ten so
+ 24	4 + 8 = 1 ten and 2 ones.
42	The 2 is written in the ones column; the 1 ten is carried to the tens column at the top.

Pupils move from 'exchanging' to 'carrying' as they move from a concrete representation to the abstract form. Allow pupils to use teaching aids until they do not need them any more.

- Later examples should include additions that pupils often find difficult, for example,

  36 + 7 where one number has no tens. (Pupils can 'count on' to check their solution to small number additions of this type.)

  25 + 50 where one number has zero in the ones column,

  25 + 5 = 30 where the answer contains zero.

Stage 3 Three-digit numbers

These are added by using the same methods and the examples should build the techniques gradually.

- Carrying to the tens only.
  For example, 134 + 357 and 222 + 339.
- Carrying to the hundreds only.
  For example, 134 + 881 and 322 + 386.
- Carrying everywhere.
  For example 134 + 287. This is demonstrated with teaching aids below.

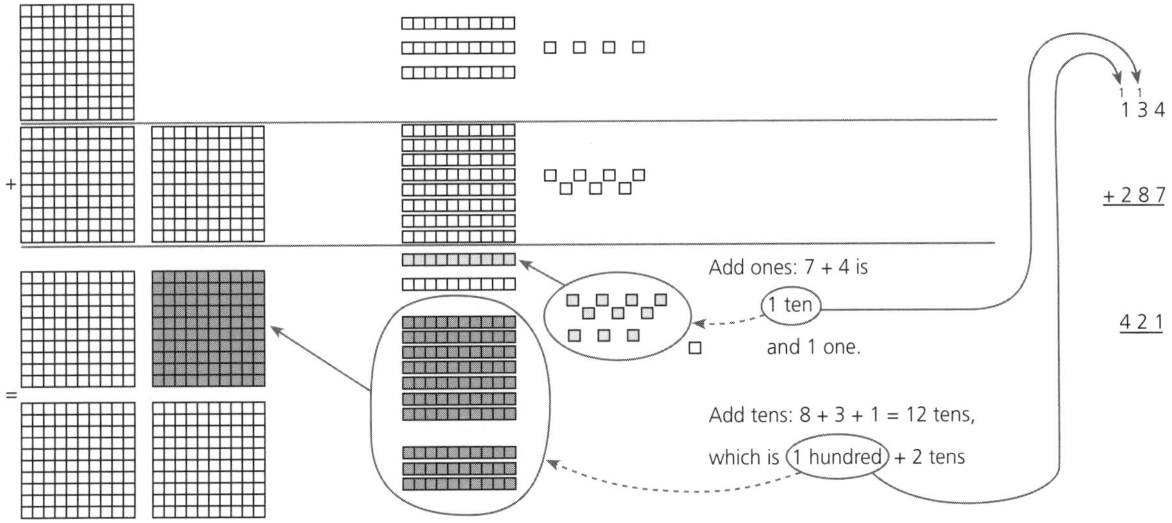

4 Larger numbers

By this stage, pupils should have understood how the algorithm works. From now on, addition of larger numbers, such as 5846 + 1283 can be explained without the use of equipment.

5 Common errors in addition

- **Number facts** Errors occur through not knowing addition facts – use remedial work (flash cards, games, etc.) to practise the addition facts.
- **Place value** The main error that pupils make is to put numbers in the wrong column (due to lack of understanding of place value).

  For example:
$$\begin{array}{r} 261 \\ +\ 17 \\ \hline \\ \hline \end{array}$$

> ***Activity***
>
> Practise how to add two-digit numbers using teaching aids by doing several examples practically, in the order presented above.
>
> Consider what you can do to correct the errors described above – without resorting to just explaining the method again!

- **Carrying errors**

(i) Forgetting the carrying figure:

```
 18
 27
 --
 35
```

(ii) Writing the wrong carrying figure:

```
  5
 18
 27
 --
 81
```

(iii) Adding carrying figure twice:

```
 ¹18
 ¹27
 ---
  55
```

(iv) Carrying when there is nothing to carry

```
 12
 27
 --
 49
```

6 An exchange game

1 Each player rolls two dice (or picks a card) and collects the number of sticks or ones indicated.

2 When a player has 10 sticks/ones they are exchanged for a bundle/long.

Score
7
6
9

Bundles/ Longs	Sticks/ Ones
	7
1	3
2	2

3 The winner is the first to have 10 bundles/longs.

4 Players can record their results as they go along in a table with headed columns. Later, the headings could be H, T, and O.

Topic 6 Basic subtraction number facts

1 Introduction

Subtraction arises in several situations. These include:

- **taking away:** I had 8 apples and I ate 3 of them.
  There are 8 – 3 = 5 left.
- **comparison:** I have 8 apples. My sister has 3 apples.
  I have 8 – 3 = 5 more apples than her.
- **inverse addition:** I have 3 cents. I want to buy an 8¢ sweet.
  I need 8¢ – 3¢ = 5¢ more.

In each case, a set has split into two subsets. Normally subtraction is introduced as 'taking away'. However, pupils should meet examples involving other kinds of subtraction.

2 The subtraction facts

Subtraction is the inverse of addition. Subtracting 5 'undoes' adding 5. Each addition fact is associated with two subtraction facts.

For example, 5 + 4 = 9 is connected to 9 – 4 = 5
and 9 – 5 = 4.

Learning the basic subtraction facts is closely related to the knowledge of the associated addition facts. Link them when teaching. The basic number facts can be discovered using teaching aids. Let children use physical materials until they feel confident enough to work only with the symbols.

3 Teaching aids

Like the addition facts, the basic subtraction number facts need to be learned by all pupils. As well as using the addition facts to find out subtraction facts, pupils should find out these facts using a variety of materials and methods including the following.

- **Small objects, people, fingers and so on**

  For example, 9 – 4 = 5 can be established by putting 9 objects (or children) in a group and then removing 4 of the objects (children) and counting the 5 objects (children) left.

  Fingers can be used when small numbers are involved.

  For example: 7 fingers take away 3 fingers leaves 4 fingers.

- **A variety of vocabulary**

  For example:
  The difference between 3 fingers and 7 fingers is 4 fingers.
  7 fingers is 3 fingers more than 4 fingers.

- **Cuisenaire rods or strips**

For example, 6 take away 4 leaves 2.

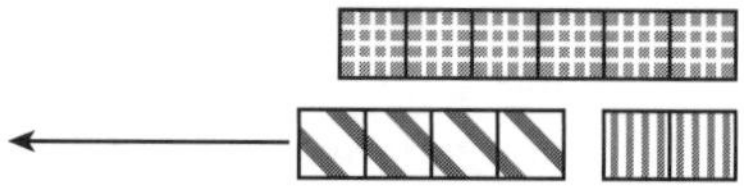

- **Partitions**

For example, the partitions of 7 can be used to show that:
6 = 7 − 1
5 = 7 − 2
4 = 7 − 3
and so on.

- **The number road**

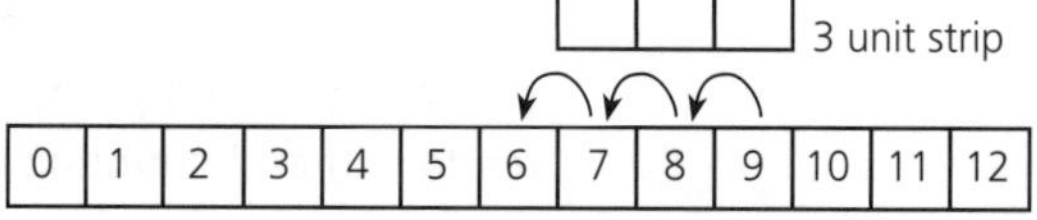

The number road can be used to do subtraction.

For example, to find out what number is 3 less than 9, count back 3 squares from 9 to 6, or put a 3 unit strip next to 9 on the road and look for its beginning.

Activity

1 Practise using Cuisenaire rods or strips.

2 Make a number road. Practise using it to teach subtraction facts.

Another way of finding 9 – 3 is to count on from 3 to 9.
A tape measure can also be used:

0	1	2	3	4	5	6	7	8	9	10	11	12

The numbers here are next to the lines.

4 Helping children to learn number facts

There are many subtraction number facts that have to be learnt and applied, without resorting to learning aids. What can be done to make this task easier?

- Many of the subtraction facts are in pairs, for example 7 – 4 = 3 and 7 – 3 = 4. Knowing this helps the pupils to remember them.
- Help the pupils to learn through practice that is fun, and use Games or Flash cards.
- Money and other practical situations can be used.

5 A subtraction game

Making twelve

This game is for two players. In each row in the table below, every number has a 'partner' with which it adds up to 12 (8 + 4, 7 + 5, and so on).

8	5	9	2	10	3	7	4
1	10	6	5	7	6	2	11
3	7	4	10	2	8	5	9
9	8	7	6	6	5	4	3

One player chooses a number and puts a blue counter on it. The other player says the number **in the same row**, which added to the first number gives twelve and puts a red counter on it. Repeat for other numbers.

Activity

Develop some activities that will help your pupils learn addition and subtraction number facts. Are there any songs that can help? Can you develop some games?

Topic 7 Subtraction algorithm

1 Introduction

When the children know the basic addition and subtraction number facts they are ready to subtract larger numbers. The order in which the different types of subtraction problems are taught should be carefully managed.

2 Subtraction problems with no 'borrowing' or 'decomposition'

It does not matter whether you start with the ones or the tens when there is no borrowing. However, get your pupils into the habit of starting with the ones, as they will need to do this when the ones digit subtracted is larger than the digit from which it is subtracted. The first example should have no zeros in it or in the answer.

For example: 59 – 32

This can be expressed as:

- Take 32 away from 59.
- How many more is 59 than 32?
- What is the difference between 32 and 59?

At first, the pupils should use the same equipment that they used for addition to help them understand the principles involved.

This is the method with the strips of paper.

$$\begin{array}{r} 59 \\ -\,32 \\ \hline 27 \\ \hline \end{array}$$

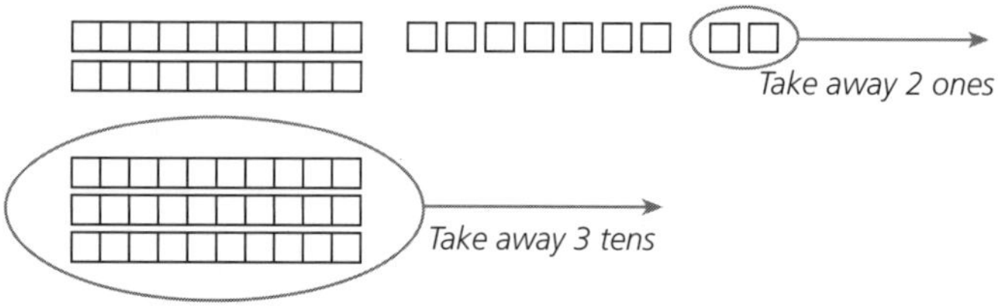

2 tens and 7 ones are left behind.

In symbols:

$$\begin{array}{r} 59 \\ -\,32 \\ \hline 27 \\ \hline \end{array}$$

In words: start with the ones: 9 take away 2 leaves 7.
Put 7 in the ones column of the answer.
Now the tens: 5 take away 3 leaves 2.
Put 2 in the tens columns of the answer.

With sticks and bundles it would be like this:

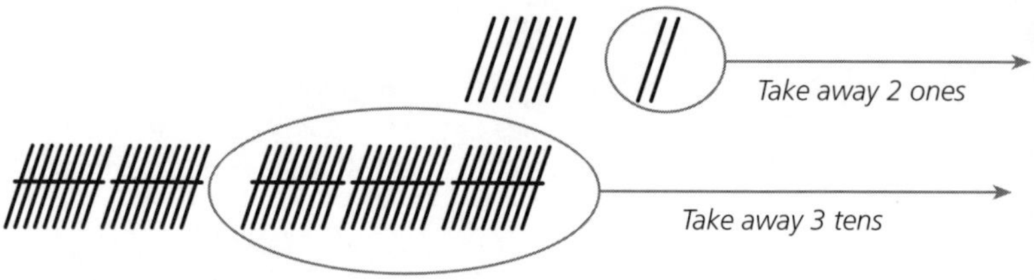

Order of examples

As with addition, the order in which the examples are taught should be carefully chosen.

(i) only tens and ones, no zeros
For example: 59 – 32   36 – 23
(ii) subtraction of ones only
For example: 56 – 5
(iii) zero in ones column
For example: 38 – 18   54 – 30
(iv) gaps/zeros in tens column of answer
For example: 38 – 33   54 – 51
(v) hundreds, tens, and ones
For example: 735 – 421
(vi) hundreds, tens, and ones, with zeros
For example: 755 – 45   483 – 81   299 – 294

3 Subtraction problems with 'borrowing' or 'decomposition'

> ***Activity***
>
> Try out these examples with the longs, bundles and on an abacus.
>
> Compare the three approaches.
>
> Why do you think that examples with zeros in them cause some pupils problems?

The term 'borrow' is misleading. The number that is used from the tens columns is not returned, so it has not been 'borrowed.' A more accurate description is to say that a ten is *decomposed* (broken down) into ten ones.

Order of examples

It is important to build up the difficulty of the examples methodically.

(i) only tens and ones, no zeros
For example: 35 – 19

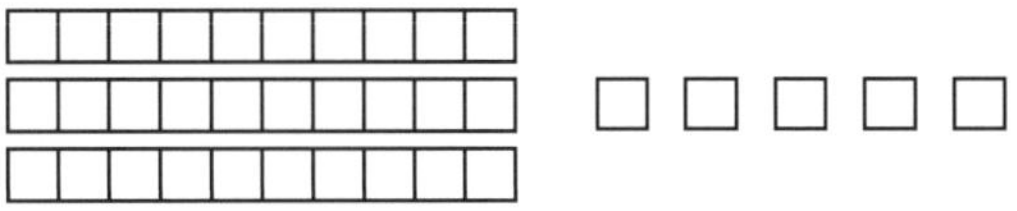

9 sticks cannot be taken from 5 sticks so decompose a bundle of ten.

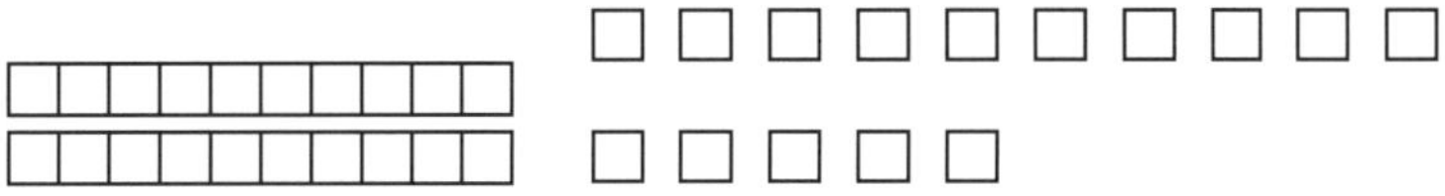

Now we can remove 9 ones from the 15 ones, and 1 ten from the 2 tens remaining.

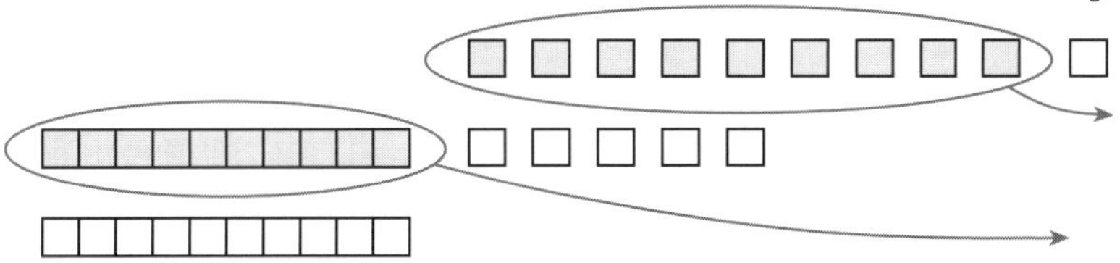

This leaves 6 ones and 1 ten.

$$\begin{array}{r} {}^{2}\not{3}\,{}^{1}5 \\ -\,1\,9 \\ \hline 1\,6 \\ \hline \end{array}$$

In symbols: Starting with the ones, 9 cannot be taken from 5, so a ten has to be broken into ones, (or decomposed). Take 1 of the tens and write it as *10 ones*, making 15 ones. (Some people write 10+ next to the 5.) Now subtract 9 ones from 15 ones, and 1 ten from 2 tens.

(ii) ones only in the bottom line
For example: 56 – 9

(iii) zeros in the ones column
For example: 30 – 18

(iv) gaps/zeros in the tens column of the result
For example: 34 – 28

(v) hundreds, tens, and ones
For example: 475 – 128 (decomposition of tens only)
435 – 261 (decomposition of hundreds only)
548 – 259 (decomposition of tens and hundreds)

```
 4 13 1
 5̸ 4̸ 8
- 2 5 9
  -----
  2 8 9
```

(vi) bigger numbers
For example: 5842 – 2673

4 Some causes of errors when subtracting

- Failure to remember subtraction facts.
- Failure to change the number in the tens column after a ten has decomposed.

```
  35
- 17
 ---
  28
```

- Faulty setting down.

```
 543
- 27
 ---

 ---
```

- Taking the smaller digits from the larger digits.

```
  35
- 16
 ---
  21
```

Activity

Some people use an *equal addition strategy* to simplify subtraction.

52 – 27 is the same as 55 – 30 (start with 3 more and take away 3 more), which it is easy to see is 25. By adding 3 to both numbers, the subtraction becomes simpler.

Similarly: 88 – 49 = 89 – 50 = 39

Gerdes asked street traders in Mozambique how they would work out 62 – 5.

None used the formal algorithmic method.

The most popular methods were: 62 – 2 = 60, 60 – 3 = 57 (59% of them)

60 – 5 = 55, 55 + 2 = 57 (29%)

These informal methods are effective.

- Should these methods be taught?
- Do you use any unusual or different methods?

Topic 8 Multiplication tables

1 Introduction

Much of what children learn in maths depends on their understanding and remembering what they have learned previously. It is very important that pupils learn basic arithmetic facts so that they can recall them quickly. It is very helpful if they can say the answer to questions like '*What is* 7×9*?*' quickly and correctly, without having to recite 1×9, 2×9, 3×9, ... first. This is because they will be doing a lot of multiplication all through their mathematical life when solving other problems.

2 Learning the multiplication facts

Stage 1 Multiplication arises in several different ways. These include:

- **repeated addition**
  For example: 1 person has 2 legs, 3 people have $2 + 2 + 2 = 6$ legs
- **equal groups** — several sets have the same number in each.
  For example: 3 children eat 2 apples each. How many apples are eaten?
- **phrases such as '... times as many/much'**
  For example: I have \$2. My brother has 3 times as much. How much does he have?
- **rate**
  For example: Sweets cost 2¢ each. How much do 3 sweets cost?

The pupils need to build up their knowledge of the multiplication tables. Some tables are easier than others to remember:

- $\times$ 10 is the easiest, $\times$ 5 comes next, then $\times$ 2, $\times$ 3, and $\times$ 4
- the hardest are $\times$ 6, $\times$ 7, $\times$ 8, and $\times$ 9.

Using the number line and repeated addition

For example, 6×3 means 3 lots of 6, or $6 + 6 + 6$.

The sum can be found by counting out 3 lots of 6 objects and counting them one by one. The result can be shown effectively on a number line:

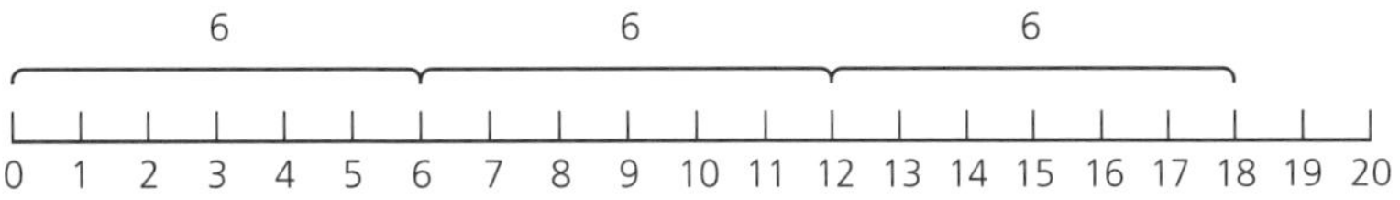

Numbers of things

Use situations associated with multiplication that can help pupils learn the multiplication facts. For example, people have five fingers on one hand. Two hands have 10 fingers, three hands have 15 fingers, and so on. Pupils can hold up their hands one by one in order to develop the 5 times table.

Here are some examples for all the numbers up to 8:

2 pairs of things	Count legs, eyes, nostrils, etc. of several people. Count wheels on some bicycles.
3 threes	Count finger segments, wheels on tricycles, sides of triangles.
4 fours	Count legs of chairs, wheels of cars, sides of rectangles and squares.
5 fives	Count fingers on hands, toes on feet.
6 sixes	Count legs on pictures of insects, sides of hexagons.
7 sevens	Count days in a week. Look at a monthly calendar to find 7×2, 7×3, 7×4, etc.
8 eights	Count legs on pictures of spiders.

Multiplying by nine

Because 9 is 1 less than 10 there is a pattern to the multiples 9, 18, 27, 36, 45, 54, 63, 72, 81. The sum of the digits is 9 in each case. Also, the number in the tens column is 1 less than the number 9 is being multiplied by.

Finger multiplication: the nine times table

The hands can be used to work out the 9 times table.

For example, 9×3.

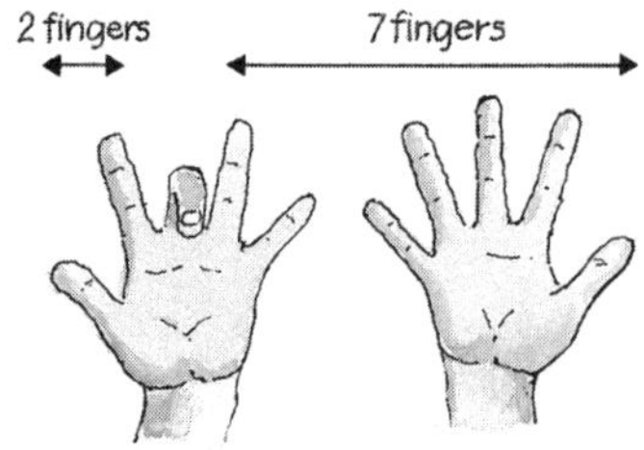

Hold up both hands. 9 is multiplied by 3 so fold the 3rd finger from the left.

There are 2 fingers to the left and 7 fingers to the right of the folded finger. The answer is 27.

Stage 2 Remembering the multiplication facts

Draw attention to the fact that if the multiplication tables are written in a 10 by 10 square, there is a symmetrical pattern about the diagonal.

×	1	2	3	4	5	6	7	8	9	10
1	1	2	3	4	5	6	7	8	9	10
2	2	4	6	8	10	12	14	16	18	20
3	3	6	9	12	15	18	21	24	27	30
4	4	8	12	16	20	24	28	32	36	40
5	5	10	15	20	25	30	35	40	45	50
6	6	12	18	24	30	36	42	48	54	60
7	7	14	21	28	35	42	49	56	63	70
8	8	16	24	32	40	48	56	64	72	80
9	9	18	27	36	45	54	63	72	81	90
10	10	20	30	40	50	60	70	80	90	100

This is because multiplication is commutative. If a pupil knows that $6 \times 9 = 54$, she also knows that $9 \times 6 = 54$. This considerably reduces the number of facts to remember, (down from 100 to 55 for tables up to 10×10).

These are:

1 × 1	1 × 2	1 × 3	1 × 4	1 × 5	1 × 6	1 × 7	1 × 8	1 × 9	1 × 10
	2 × 2	2 × 3	2 × 4	2 × 5	2 × 6	2 × 7	2 × 8	2 × 9	2 × 10
		3 × 3	3 × 4	3 × 5	3 × 6	3 × 7	3 × 8	3 × 9	3 × 10
			4 × 4	4 × 5	4 × 6	4 × 7	4 × 8	4 × 9	4 × 10
				5 × 5	5 × 6	5 × 7	5 × 8	5 × 9	5 × 10
					6 × 6	6 × 7	6 × 8	6 × 9	6 × 10
						7 × 7	7 × 8	7 × 9	7 × 10
							8 × 8	8 × 9	8 × 10
								9 × 9	9 × 10
									10 × 10

The difficult facts to remember are usually the ones near the bottom of this table. Here are some strategies for helping pupils to know the multiplication facts well.

- Pupil writes out multiplication table.
  Pupil crosses through those lines that he or she is familiar with.
  Pupil learns the rest.
- **'Mini-tables'** for completion. They are set out mixed up, so the pupil cannot complete them by recitation.

×	8	5	3
6		30	
8			
7			
2			

×	7	2	9
2			
3			
4			
6			

- **Multiplication 'arithmogons'** for completion. If you multiply the numbers in two circles together, you get the numbers in the squares between them.
  For example:

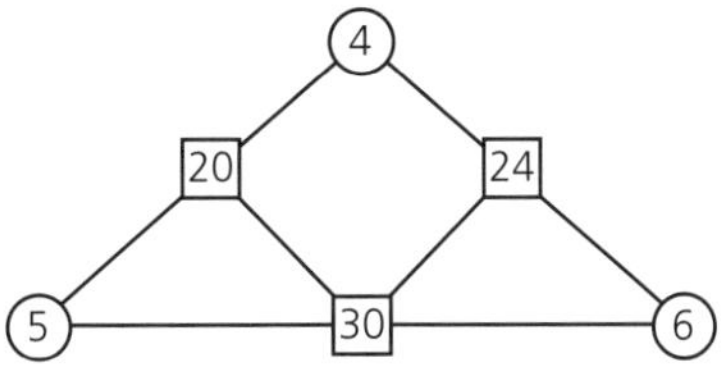

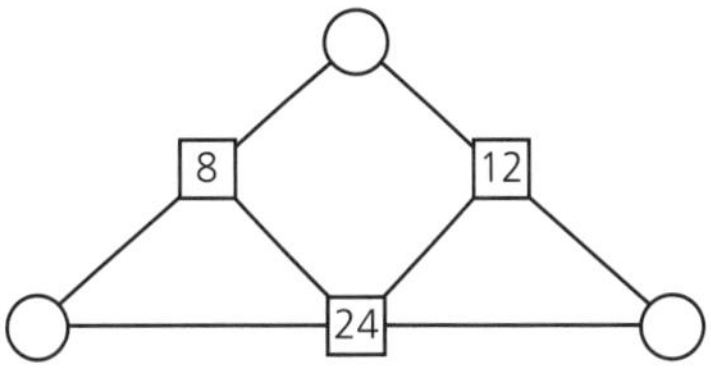

- **Games involving the tables**
  Games like Flash cards, number pairs, dominoes are very good for practising multiplication number facts.

Activity

Look for patterns in the hundred square of tables:

- Shade all the even numbers red, and all the multiples of 5 blue. What do you notice?
- Repeat for multiples of 3 and 4, and multiples of 4 and 8.
- Repeat for multiples of 9 and 3, and multiples of 2 and 3.
- What do you notice?

Topic 9 Multiplication algorithm

1 Introduction

The multiplication algorithm is complex, and teaching it involves more stages than the addition algorithm. It is important that pupils have learnt the multiplication facts for numbers up to 10.

2 Order of examples

The order of introducing examples is important. Begin with products for which no *carrying* (also called *regrouping*) is required, and with small numbers. The following order will be useful to follow.

Stage 1 Multiply a multiple of 10 by a 1-digit number.
For example: 20×4

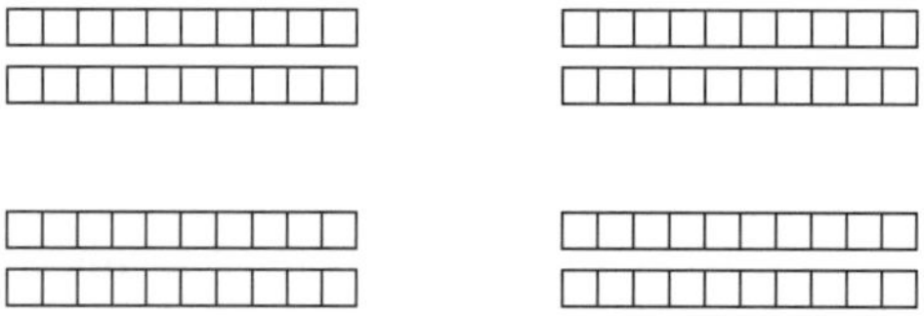

In symbols:

$$\begin{array}{r} 20 \\ \times\ 4 \\ \hline 80 \\ \hline \end{array}$$

Then use examples with answers greater than 100, for example $30 \times 8 = 240$.

Stage 2 Multiply any 2-digit number by a 1-digit number without carrying.
For example: 12×3

3 groups of 2 make 6
3 groups of 10 make 30

$$\begin{array}{r} 12 \\ \times 3 \\ \hline 6 \\ 30 \\ \hline 36 \\ \hline \end{array}$$

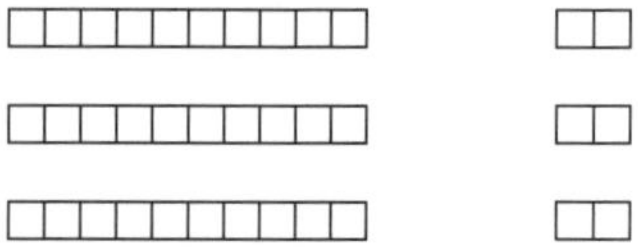

Stage 3 Multiply any 2-digit number by a 1-digit number with carrying.
For example: 16×3

3 groups of 6 make 18
3 groups of 10 make 30

$$\begin{array}{r} 16 \\ \times 3 \\ \hline 18 \\ 30 \\ \hline 48 \\ \hline \end{array}$$

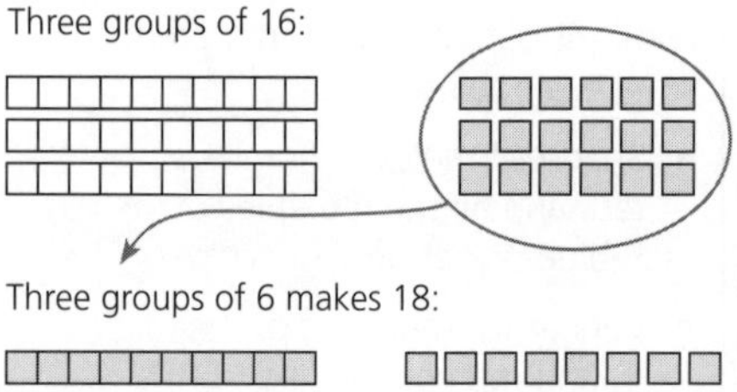

More briefly:

$$\begin{array}{r} {}^{1}16 \\ \times 3 \\ \hline 48 \\ \hline \end{array}$$

The longer form is better to start with.
The shorter form can come later.

Stage 4 Multiply larger numbers by a 1-digit number.

- 3 digits × 1 digit
  - (a) with no carrying, for example 123 × 3
  - (b) carrying tens, for example 318 × 3
  - (c) carrying hundreds, for example 152 × 3
  - (d) carrying both, for example 318 × 8
  - (e) with zeros, for example 280 × 3
- more than 3 digits × 1 digit

Stages 1 to 4 are commonly called *short multiplication*. *Long multiplication* is considered in stages 5 to 7.

Stage 5 Multiply multiples of 10 by 10 and by multiples of 10, for example 30 × 20.

Stage 6 Multiply by a 2-digit number.
(a) with no carrying, for example 23 × 13, 12 × 14

	14
	x 12
2 groups of 4	8
2 groups of 10	20
10 groups of 4	40
10 groups of 10	100
	168

Twelve groups of 14:

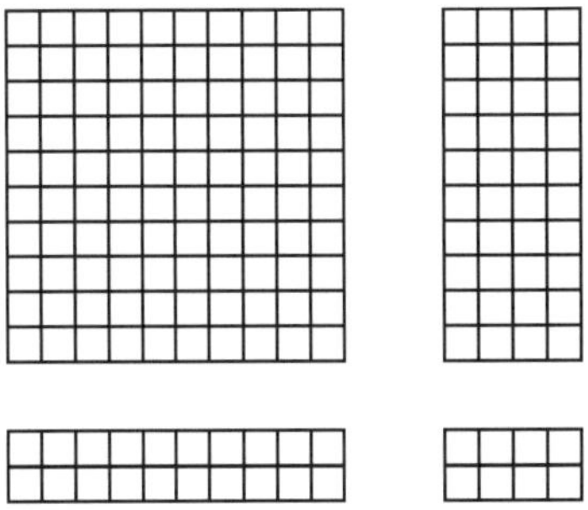

Short method:	14
	× 12
Two groups of 14	28
Ten groups of 14	140
	168

(b) with carrying, for example 38 × 43

Using teaching aids would be very tedious for calculations such as this one. Pupils should no longer need concrete materials when they attempt problems like these.

Stage 7 Multiply by any whole number.

In full:

	38
	× 43
3 × 8	24
3 × 30	90
40 × 8	320
40 × 30	1200
	1634

More briefly:

	38
	× 43
3 × 38	114
40 × 38	1520
	1634

3 Checking answers – estimates

It is a good idea for pupils to estimate the answer before attempting each calculation. This helps to avoid careless mistakes. A good way to make estimations is by making a one-significant-figure approximation of the numbers.

For example, 32×48 is approximately $30 \times 50 = 1500$. (The exact answer is 1536.)

Activity

1 Show 14×17 using apparatus.

2 Put these problems in the order in which you would teach them:

34×23	5×7	23×3	324×43	56×7
47×10	29×17	426×6	40×7	38×30

Topic 10 Division facts

1 Introduction

Division is the inverse of multiplication and each multiplication fact is associated with two division facts.

For example, $6 \times 3 = 18$ so $18 \div 3 = 6$ and $18 \div 6 = 3$.

2 Learning the division facts

All the division facts can be derived from the multiplication tables. Indeed the multiplication grid can be used to find them.

For example: To find $28 \div 7$ from the grid, go down from **7** in the top line to **28** and read across to the number at the left, which is **4**. So $28 \div 7 = 4$.

×	1	2	3	4	5	6	**7**	8	9	10
1	1	2	3	4	5	6	7	8	9	10
2	2	4	6	8	10	12	14	16	18	20
3	3	6	9	12	15	18	21	24	27	30
4	4	8	12	16	20	24	**28**	32	36	40
5	5	10	15	20	24	28	32	36	45	50
6	6	12	18	24	30	36	42	48	54	60
7	7	14	21	28	35	42	49	56	63	70
8	8	16	24	32	40	48	56	64	72	80
9	9	18	27	36	45	54	63	72	81	90
10	10	20	30	40	50	60	70	80	90	100

The division facts can also be established without using the multiplication table by using the *meaning* of division. Division is used to describe different types of action which include:

- **using the number line**
  This is very effective for repeated subtraction work. For example, $20 \div 4$ can be worked out by counting how many times we can move 4 steps back from 20 to 0. The answer is 5.

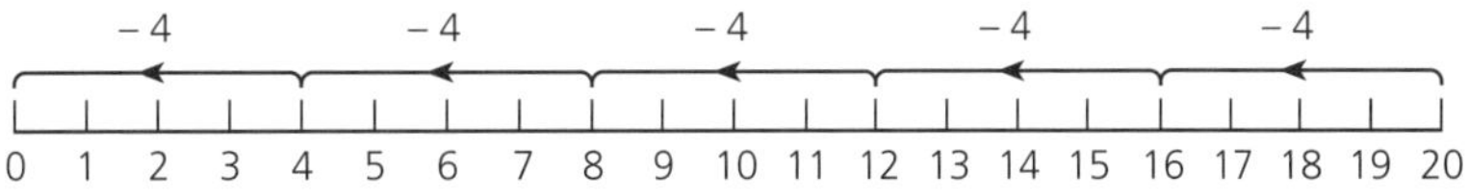

- **sharing or partitioning**
  For example, 6 children share 18 sweets. The sweets are distributed one by one to each child until all the sweets have been shared out. Each child gets 3 sweets.
  So 18 ÷ 6 = 3.
- **grouping or repeated subtraction**
  For example, 18 children are to go on a school trip. If 6 children can go in one car, how many cars are required? To find the answer we take groups of 6 away from 18 repeatedly until none are left. This we can do 3 times. We also write 18 ÷ 6 = 3

When sharing we know the number of groups to form (6 children) but do not know the number in each group. When grouping we know the number in each group (6 children per car) but not the number of groups. Diagrammatically:

	Number in group		Number of groups		
Sharing	?	×	6	=	18
Grouping	6	×	?	=	18

To work out 36 ÷ 9 by sharing, take 36 items and put them one by one into 9 piles. When the task is finished, how many are in each pile? Answer: 4.
To work out 36 ÷ 9 by grouping take 36 items and remove them 9 at a time. How many times can this be done? Answer: 4.

3 Remembering the division facts

Pupils need practice in using the division facts they have discovered in order to remember them. There are various ways this can be done.

Division Sun (*sometimes called a 'Table Spider'*)
A number is put in a small circle and lines radiate out from the circle. At the end of each line is a multiple of the number inside the circle (see diagram below for a Division Sun for 7). These multiples are in a random order.

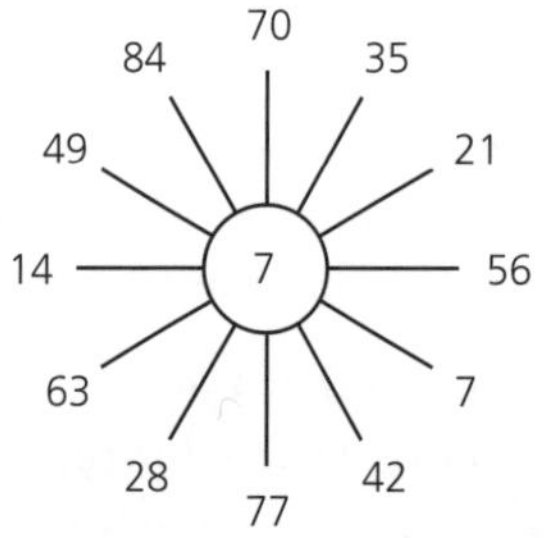

The teacher points to a number and asks how many 7s are in it, i.e. how many times can 7 be subtracted from the number? The pupils can copy the diagram into their exercise books.

> ***Activity***
>
> Develop some activities that can help your children learn the division facts in interesting and fun ways.
>
> Are there any rhymes that will help?

Flash cards Prepare flash cards with division fact questions on them (e.g. 54 ÷ 9) that are large enough for all pupils to be able to see them clearly. Ask pupils at random around the class division fact questions for a few minutes at the start of each lesson.

Topic 11 Long division

1 Introduction

- **18 ÷ 6 = 3** can be written out as a repeated subtraction:

Write $6\overline{)18}$

6	Take away 1 six	1 × 6
12		
6	Take away another six	1 × 6
6		
6	Take away a 3rd six	1 × 6
0	leaving a remainder of 0.	3 × 6

Any division problem, such as 96 ÷ 3, could be worked out by taking the divisor, 3, away from 96 repeatedly until nothing is left. This repeated subtraction takes place 32 times, so 96 ÷ 3 = 32. This would be very tedious. A shorter method is needed. Before looking at conventional long division, here is a slightly different method:

- **92 ÷ 4** Instead of taking 4 away from 92 repeatedly until nothing is left, larger multiples of 4 could be taken away:

Write $4\overline{)92}$

40	Take away 10 fours	10 × 4
52		
40	Take away another 10 fours	10 × 4
12		
12	Take away 3 fours	3 × 4
0	leaving a remainder of 0.	23 × 4

So 92 ÷ 4 = 23.

- **784 ÷ 14**

Write $14\overline{)784}$

280	Take away 20 fourteens	20 × 14
504		
280	Take away another 20 fourteens	20 × 14
224		
140	Take away 10 fourteens	10 × 14
84		
56	Take away 4 fourteens	4 × 14
28		
28	Take away 2 fourteens	2 × 14
0	leaving a remainder of 0.	56 × 14

This method of division does involve repeated subtraction but in each line a large number is taken away. The process is shorter, but requires more understanding.

2 Order of long division examples

As in most areas of maths, the order of examples when teaching long division is most important. A possible order is as follows.

Stage 1 Divide a 2-digit number by a 1-digit number (also known as *short division*). Increase the difficulties slowly. Start with questions that work out exactly first. However, remainders that are zero **within** the calculation cause difficulties, so provide questions with remainders and avoid the difficulties of empty spaces.

- Start by using simple divisions such as $6\overline{)54}$, whose answer is a basic number fact, and write them in the 'long division' form.
- Use exact multiples which will give remainders in tens column only, e.g. 72 ÷ 3.
- Progress to divisions which give a remainder in the quotient (answer), e.g. 76 ÷ 5 and 79 ÷ 7.
- Introduce problems that will have zero in the ones column of the quotient (answer), e.g. 62 ÷ 3.

The method of long division

This involves repeated subtraction. In fact, the procedure is to take away the largest possible number each time and shorten the procedure as much as possible. The procedure can be illustrated by the use of the base-ten equipment used for the other operations.

Example 1 **72 ÷ 4**

1 8	How many times does 4 (tens) go into 7 tens? (How many times can you take 4 tens away from 7 tens?) Once.
4)7 2	7 ÷ 4 = 1 remainder 3 tens.
4 0	Take away 10 fours.
3 2	How many times does 4 go into 32? 32 ÷ 4 = 8
3 2	Take away 8 fours, leaving a remainder of 0.
0	

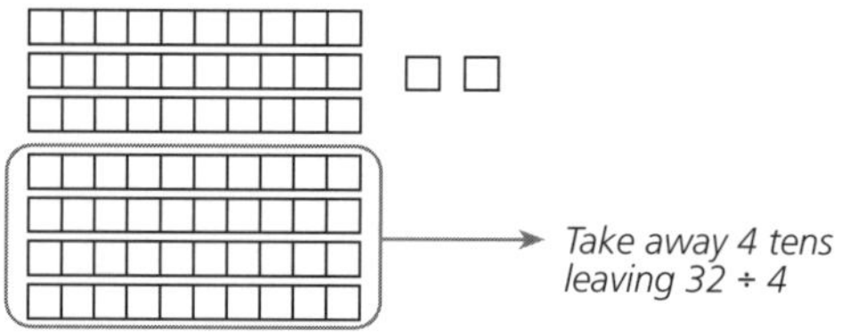

This example leaves no remainder.

Example 2 **188 ÷ 4**

3 1	
6)1 8 8	6 does not go into 1 (hundred), but 6 goes 3 times into 18 (tens). 18 ÷ 6 = 3.
1 8	Take away 3 sixes, leaving 0 tens.
0 8	Now look at the 8 ones.
6	8 ÷ 6 = **1** remainder 2.
2	Take away 1 six.

So 188 ÷ 6 = 31 remainder 2.

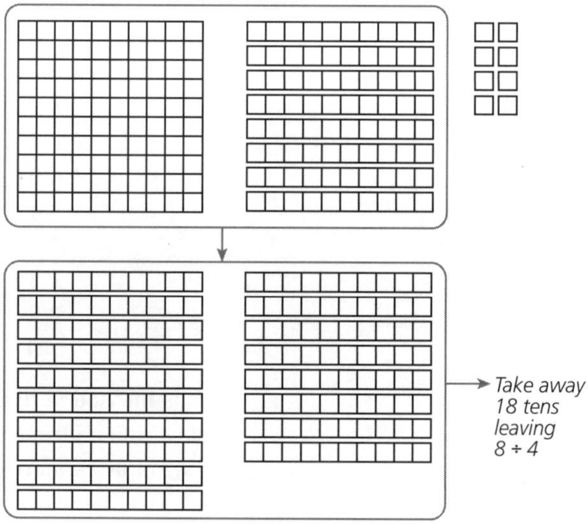

Stage 2 Divide a 3-digit number by a 1-digit number.

- Start with problems with no remainders, e.g. 682 ÷ 2.
- Show how to deal with remainders in all three columns, e.g. 716 ÷ 3.
- Progress to remainders in 2 columns, e.g. 976 ÷ 8, 849 ÷ 7, 697 ÷ 6.
- Then have examples with remainders in 1 column, e.g. 855 ÷ 5, 791 ÷ 7, 487 ÷ 4.
- Introduce examples with zeros in the quotient (answer), e.g. 825 ÷ 4, 652 ÷ 5.
- Finally use examples with zeros in dividend (question), e.g. 805 ÷ 4.

Stage 3 Divide by simple 2-digit numbers such as 21, 31, etc. These are easier than dividing by 28, 38, etc. because the multiples of 21, 31, etc. are easier to work out.

Stage 4 Divide by any number, for example 702 ÷ 18. To work this out the 18 times table is needed. It is a good idea to write out the multiples of 18 before doing the calculation:

1 × 18 = 18
2 × 18 = 36
3 × 18 = 54
4 × 18 = 72
5 × 18 = 90
6 × 18 = 108
7 × 18 = 126
8 × 18 = 144
9 × 18 = 162
10 × 18 = 180

3 9	
18) 7 0 2	18 does not divide into 7 but it does divide into 70. How many times does 18 go into 70? 18 × 3 = 54
5 4	
1 6 2	Take away 54. This leaves 16 tens; 16 is not divisible by 18, so add on the 2 ones.
1 6 2	How many times does 18 go into 162? 18 × 9 = 162
0	Take away 9 eighteens, leaving 0.

So 702 ÷ 18 = 39.

When the technique of division is understood, children should be allowed to use calculators for divisions with large numbers.

3 Difficulties with long division

- Many children have difficulty choosing the largest relevant multiple of the divisor. To find 702 ÷ 18, they need to know the multiples of 18. It is helpful to write them out at the start of the problem.
- A very common error is to omit zeros from the quotient (answer). For example, many children will write:

2 8	
18) 3 7 4 4	18 × 2 = 36.
3 6	Take away 36. This leaves 1 (hundred), not divisible by 18, so add on the 4 tens (bring down the 4). 14 is not divisible by 18 so bring down the next 4.
1 4 4	
1 4 4	18 × 8 = 144
0	Take away 144, leaving 0.

Many children forget to write in a 0 (as above), to record that 14 (tens) are not divisible by 18, and write the answer 28, when it should be 208.

- Pupils forget the basic multiplication and division number facts.
- Pupils make errors with remainders.
- Pupils forget to add on (bring down) the next number.

4 Checking

It is always a good idea to check a division answer by multiplying back again. If pupils have found, for example, that $1456 \div 26 = 36$, they should check if $26 \times 36 = 1456$. When they find the multiplication gives a result of 936, they are immediately alerted to realising they have made an error in their division. If a pupil has found that $3744 \div 18 = 38$ the teacher can show this is wrong by working out 38×18, and finding 684.

5 Estimation

Estimation is another quick way of checking.
An estimate for $3744 \div 18$ is $4000 \div 20 = 200$, which shows that the answer of 38 is far too small.

Activity

Prepare teaching aids that pupils can fill in to practise long division of three-figure numbers by a single digit. They are provided with an outline of the division to fill in.

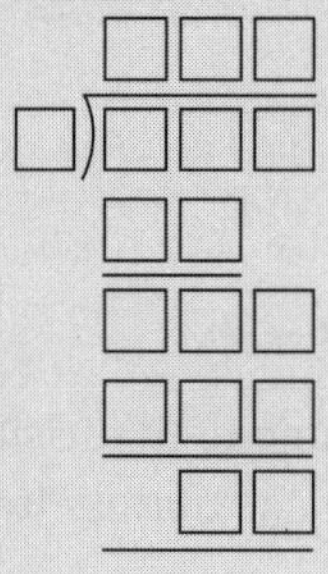

So □□□ ÷ □ = □□□ remainder □□

Topic 12 Introducing fractions

1 Introduction

The word 'fraction' comes from the same source as 'fracture' and refers to breaking into small pieces. A fraction expresses the relationship between **a whole thing and a part of that thing.**

What is the 'whole thing'? It does not have to be 1 (a single thing). It can be more than one. We can find a fraction of:

- a single object (such as $\frac{3}{4}$ of a circle)
- a number of objects (such as $\frac{1}{5}$ of 20 sweets)
- a multiple of a number of objects (such as $\frac{2}{3}$ of 6 boxes of 20 toys)
- part of an object (such as $\frac{3}{4}$ of a half of a cake).

Young children find it difficult to think of a collection of things as a single entity. But there are examples they will be familiar with, such as their family, school, class or team.

Children need to make and manipulate practical models of fractions. They are too young to think abstractly.

2 A note about notation and pronunciation

Someone once said that 'It is the notions not the notation that matters.' Fraction notation is a little tricky. A fraction can be written in several ways, for example fifteen seventeenths can be written as 15/17 or ${}^{15}\!/_{17}$ or $\frac{15}{17}$. The notation is interchangeable. It must be emphasised that two numbers are being used to describe one amount, a fraction, which is not found by counting. When it comes to naming the fraction, it is much better to say 'fifteen seventeenths' than 'fifteen over seventeen,' even though it is a bit harder to say. This is because it makes manipulating the fractions easier. It is easy to see that 'three fifths < four fifths', but 'three over five < four over five' is not at all obvious. It is easy to see that 'three fifths + one fifth = four fifths', just as 'three toys + one toy = four toys.' However, 'three over five + one over five = four over five' is not obvious.

Begin by naming fractions (e.g. two fifths) and only use the symbols later on.

3 Models of fractions

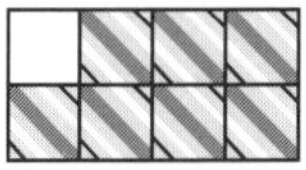

Pupils are often introduced to fractions by folding a piece of paper and shading a part of it.
The fraction shaded in the diagram is $\frac{7}{8}$ of the paper.
8 is the number of equal parts of the paper, 7 is the number of parts that have been shaded. This is just one model (an easy one to prepare) for fractions that can be used when teaching. It is a model in which a **region** has been shaded.

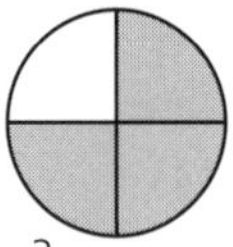

$\frac{3}{4}$ of a pie

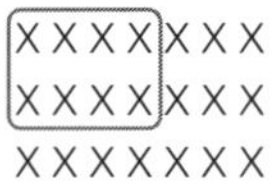

$\frac{8}{21}$ of the region is inside the string

These are some more examples of fractions of regions.

The models can also be **linear**.
For example, $\frac{5}{7}$ of this rod is shaded.

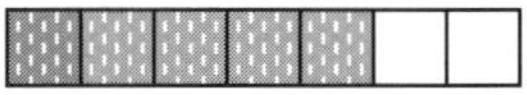

It is important to illustrate the idea of a fraction as being a **part of a number of objects**. For example, in the diagram there are 10 blocks and $\frac{7}{10}$ of them have been shaded.

4 Fractions as a part of one thing

A simple way to introduce the meaning of fractions is through the folding and shading of pieces of paper. Pupils should be given the opportunity to do plenty of this for themselves.

Emphasise that for us to be able to name the parts (fractions), they must all be equal.

For example, three quarters, $\frac{3}{4}$ is

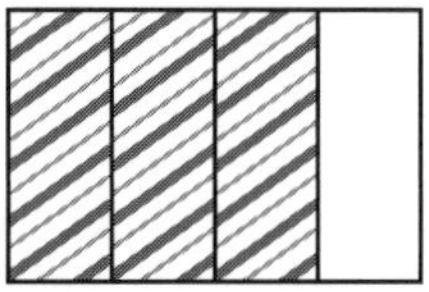

not

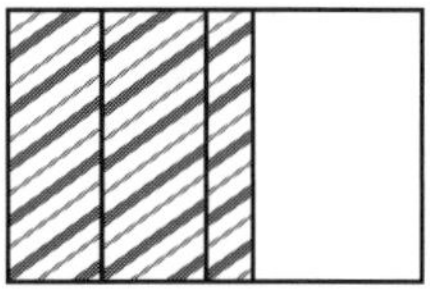

This shape is in two equal parts.
It is in two halves.

Fold a strip of paper lengthwise so that it is in eight equal parts (eighths).

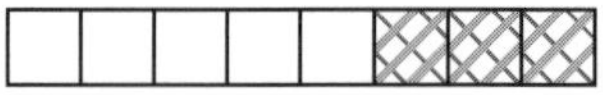

The shaded part is *three-eighths* of the rectangle. We write this as $\frac{3}{8}$.

Two-thirds (written $\frac{2}{3}$) of this shape is shaded.

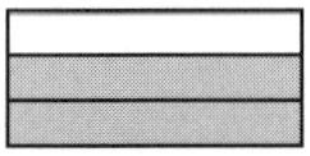

Cut out paper circles and ask the pupils to fold them into four equal parts. Shade 3 of them as in the diagram. Three-quarters (written $\frac{3}{4}$) of this shape is shaded.

These and other examples can be used to introduce the main features of fractions. Start with regions with only one part shaded, then move on to regions with several parts shaded. Show how, for example, $\frac{5}{7}$ means five-sevenths, five of seven **equal** parts. In each case, the number on the bottom, called the **denominator**, is the number of equal parts the whole thing has been divided into. The number on the top, called the **numerator**, is the number of the parts in the fraction. Some writers do not like these difficult terms. They prefer *top number* and *bottom number*.

Give pupils plenty of practice folding paper and shading parts of shapes. Make sure that they name the fractions of the folded paper that they have shaded.

5 Fractions as a part of a set of things

Talk about words used for a group, such as 'family', 'school', 'class', 'team' or 'nation'.

Suppose there are 8 pupils (5 boys and 3 girls) in a team. Ask 8 pupils to stand up to represent the team. Say that each pupil is $\frac{1}{8}$ of the team. Ask how many are boys. Tell the class that $\frac{5}{8}$ of the team is boys.

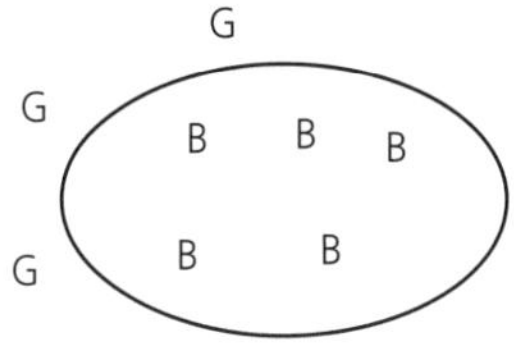

Illustrate this with a picture. Repeat with other examples.

Bottle tops are a very good resource for this. Suppose you have 12 bottle tops and 7 of them red, then $\frac{7}{12}$ of the tops are red tops. Give pupils bottle tops, and ask them to pick out different fractions.

6 Showing that $\frac{n}{m}$ is the same as $n \div m$

Show each of 3 rectangles as 8 squares. There are 24 small squares.

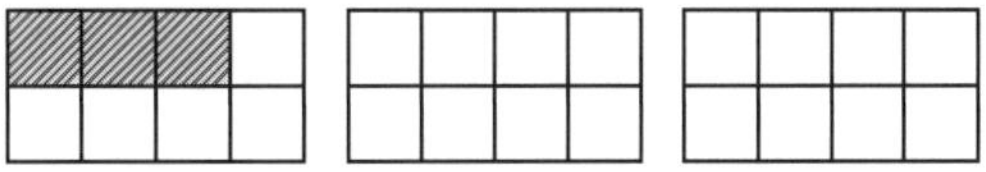

3 rectangles (or 24 squares) $\div$ **8 = 3 squares**.
Shade these 3 squares. We can see **3 squares =** $\frac{3}{8}$ **of one rectangle**.
So $\frac{3}{8}$ of a rectangle = 3 rectangles $\div$ 8.
Repeat for other fractions.

7 Number lines

Fractions can be shown on a number line. The fractions on the right are a larger amount than those on the left. For example, the number line for eighths:

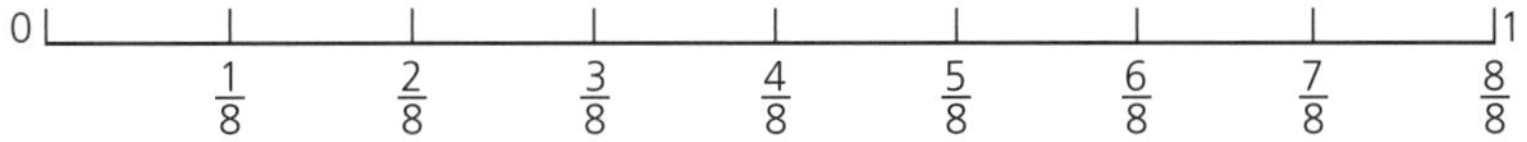

Activity

- What are the words for fractions in your local languages? Are they easy to use?
- Is it easier to name fractions such as quarters, fifths and so on in English or in the local language?
- Prepare teaching materials for the various models of fractions. Discuss which ones you find easiest to use.

Topic 13 Equivalent fractions

1 Introduction

Equivalent fractions are important in work on fractions — comparing them, adding and subtracting them, changing them into decimals and percentages. Equivalent fractions can be introduced practically, by shading and folding pieces of paper.

Two equivalent fractions are fractions that are equal but are written in different ways. It is, perhaps, a pity that they are not just simply called 'equal fractions'.

2 Introducing equivalent fractions

Shaded and folded paper is very effective for introducing equivalent fractions.

- Fold a piece of paper into 3 equal parts and shade $\frac{2}{3}$.
- Fold the paper again, in 2 equal parts and ask how much is shaded. $\frac{4}{6}$.
- Is the same amount shaded? Yes, so $\frac{2}{3} = \frac{4}{6}$.

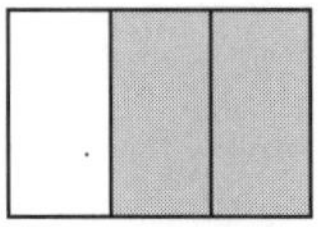

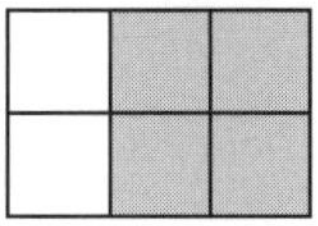

$\frac{2}{3} = \frac{4}{6} = \frac{8}{12}$

- Fold the paper in 2 equal parts again and ask how much is shaded. $\frac{8}{12}$.
- Is the same amount shaded? Yes, so $\frac{2}{3} = \frac{4}{6} = \frac{8}{12}$.

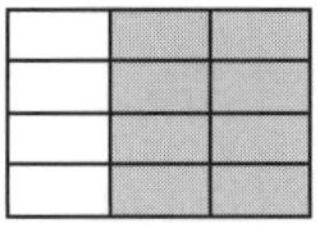

Let the pupils do a few more examples. For example they could find that $\frac{5}{8} = \frac{10}{16} = \frac{20}{32}$. Tell the pupils that fractions that are equal to each other are called **equivalent.**
Some will notice that $\frac{5}{8}$ can be changed to an equivalent fraction by **multiplying** its numerator and denominator by the same number:

$$\frac{5}{8} = \frac{5 \times 2}{8 \times 2} = \frac{10}{16}.$$

Give the pupils further practice at finding equivalent fractions in this way.

What happens if you reverse the process? $\frac{10}{16}$ can be changed to an equivalent fraction by **dividing** its numerator and denominator by the same number:

$$\frac{10}{16} = \frac{10 \div 2}{16 \div 2} = \frac{5}{8}.$$

This process, which reduces the size of the numbers in the fraction, without changing the value of the fraction, is called **simplifying**.

3 The fraction chart

This is an effective way of displaying equivalent fractions.

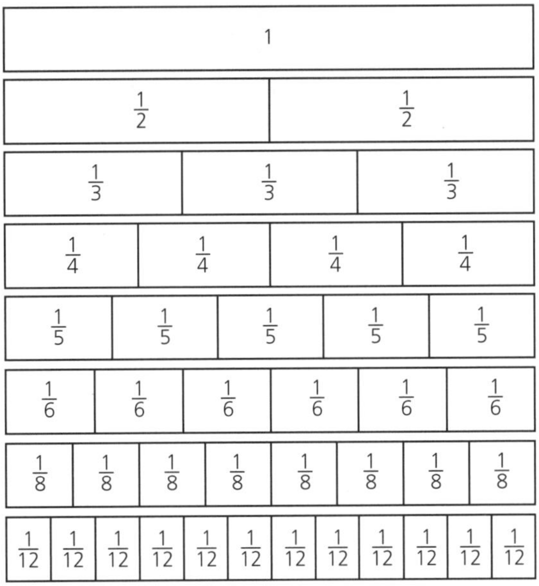

> ***Activity***
>
> Make a fraction chart for use in the classroom. Make it 60 cm wide.
>
> Work out how long each fraction block should be. Make it on card so that it keeps well. The chart above does not include 7ths, 9ths, 10ths or 11ths. You could include them in your chart. (10ths are useful when considering decimals.)

Looking down the middle of the chart, it can be seen that
$\frac{1}{2} = \frac{2}{4} = \frac{3}{6} = \frac{4}{8} = \frac{6}{12}$.

This diagram could be copied with Cuisenaire rods (see Section B, Topic 4).
For example, use 1 ten as 'one', five as $\frac{1}{2}$, two as $\frac{1}{5}$ and one as $\frac{1}{10}$.
2 ones and 1 two are the same length: 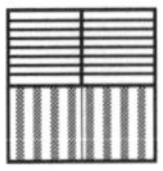 $\frac{2}{10} = \frac{1}{5}$

4 Number lines

Equivalent fractions can be shown on number lines. For example:

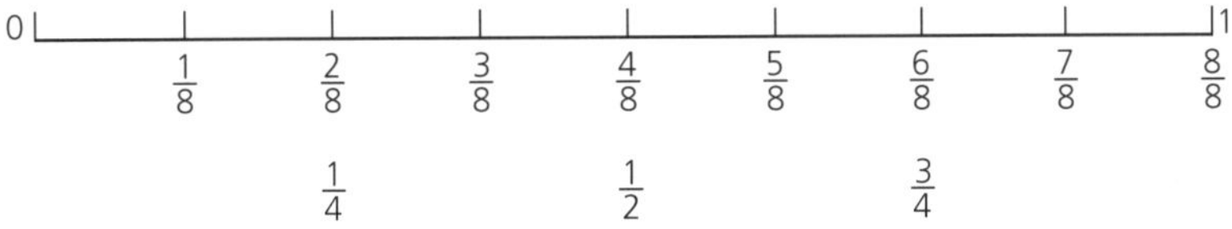

5 A difficult fraction problem

A common question to be asked is to fill in the spaces:

$$\frac{6}{7} = \frac{\ }{14} = \frac{24}{\ } = \frac{\ }{42} = \frac{66}{\ } = \frac{\ }{84} \ldots\ldots$$

The first two spaces are easy to fill: multiply by 2:

$$\frac{6}{7} = \frac{12}{14} = \frac{24}{28} = \frac{\ }{42} = \frac{66}{\ } = \frac{\ }{84} \ldots\ldots$$

The next space is more difficult. 42 is not a multiple of 28.
A common error is to see that the denominator 42 is 14 more than

the denominator 28 so next numerator must be 14 more than numerator 24 which would be 38. This error is repeated for the rest of the line and the pupil may write the following:

$$\frac{6}{7} = \frac{12}{14} = \frac{24}{28} = \frac{38}{42} = \frac{66}{70} = \frac{80}{84}\ldots\ldots$$

which, of course, is wrong. Suggest to the pupils they consider each fraction separately and then complete the answer:

$\frac{6}{7} = \frac{12}{14}$

$\frac{6}{7} = \frac{24}{}$ Multiply by 4 so $\frac{6}{7} = \frac{24}{28}$.

$\frac{6}{7} = \frac{}{42}$ Multiply by 6 so $\frac{6}{7} = \frac{36}{42}$.

$\frac{6}{7} = \frac{66}{}$ Multiply by 11 so $\frac{6}{7} = \frac{66}{77}$.

$\frac{6}{7} = \frac{}{84}$ Multiply by 12 so $\frac{6}{7} = \frac{72}{84}$.

Put them together for the answer:

$$\frac{6}{7} = \frac{12}{14} = \frac{24}{28} = \frac{36}{42} = \frac{66}{77} = \frac{72}{84}.$$

6 Comparing fractions

Fractions with the same denominator

It can be seen that by comparing two shaded rectangles that five eighths **is greater than** three eighths, $\frac{5}{8} > \frac{3}{8}$, and one quarter **is less than** three quarters, $\frac{1}{4} < \frac{3}{4}$. Other models can also be used to demonstrate this:

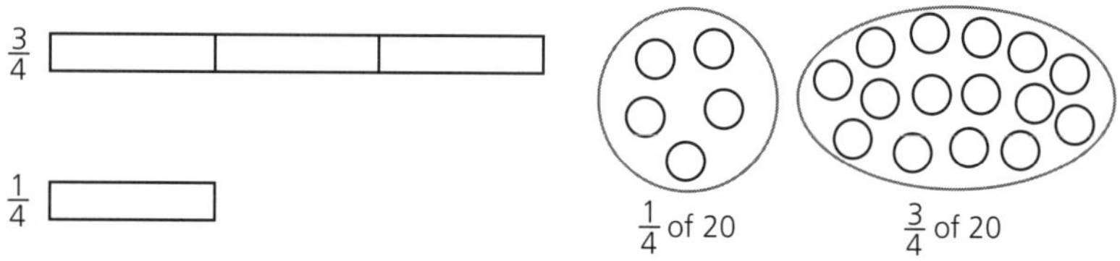

If denominators are equal, the fraction with the greater numerator is larger.

Fractions with the same numerator

Look at fractions such as $\frac{1}{4}$, $\frac{1}{12}$ and $\frac{1}{3}$. We could compare them by changing them to equivalent fractions with the same denominator ($\frac{3}{12}$, $\frac{1}{12}$ and $\frac{4}{12}$). However, it is simpler, and quicker, to think of sharing. The more people sharing, the less each person gets. So $\frac{1}{4} < \frac{1}{3}$ because 4 people sharing each get less than 3 people sharing. This can also be confirmed by looking at the size of $\frac{1}{2}$, $\frac{1}{3}$, and so on in the fraction chart.

Fractions with different numerators and denominators

Let us consider $\frac{7}{12}$ and $\frac{5}{8}$.

First look at equivalent fractions:

$$\frac{7}{12} = \frac{14}{24} = \frac{21}{36} = \frac{28}{48} = \ldots\ldots$$

$$\text{and } \frac{5}{8} = \frac{10}{16} = \frac{15}{24} = \frac{20}{32} = \ldots\ldots$$

Notice both lists have a fraction with a denominator of 24. This is called a **common denominator**.

Compare the fractions which have the common denominator:

$$\frac{7}{12} = \frac{14}{24} < \frac{15}{24} = \frac{5}{8}$$

Alternatively: notice that the denominators are 12 and 8 and multiply the top and bottom of the first fraction by 8, and the top and bottom of the second fraction by 12.

$$\text{This gives: } \frac{7}{12} = \frac{7 \times 8}{12 \times 8} = \frac{56}{96} \text{ and } \frac{5}{8} = \frac{5 \times 12}{8 \times 12} = \frac{60}{96}$$

$$\text{so } \frac{7}{12} = \frac{56}{96} < \frac{60}{96} = \frac{5}{8}.$$

(You do not need to use the lowest common denominator because you are only interested in which fraction is larger.)

Activity

Compare the two approaches given above.

Consider which you prefer, and why.

Does your preference depend on the numbers involved?

Topic 14 Adding and subtracting fractions

1 Introduction

After the pupils have understood equivalent fractions, they are ready to add fractions. Addition of fractions should start with the addition of fractions with the same denominator. Equivalent fractions are then used to apply this simple method to adding fractions with different denominators. As in much of maths teaching, it is important to grade the examples carefully.

2 Adding fractions with same the denominator

For example, consider $\frac{1}{5} + \frac{2}{5}$.

Show $\frac{1}{5}$ as part of a circle or rectangle and show $\frac{2}{5}$ as part of the same circle or rectangle.

How much is shaded altogether? $\frac{3}{5}$.

Say, 'one fifth plus two fifths'. It is logical to conclude this will be three fifths, just like one apple plus two apples is three apples.

Two important points:

(i) Tell your pupils to say fractions in this way, using the fraction words: thirds, quarters, fifths, and sixths, etc. It helps them to understand what is happening. Also, they are less likely to think that one over five plus two over five is three over ten!

(ii) Use one diagram not two.

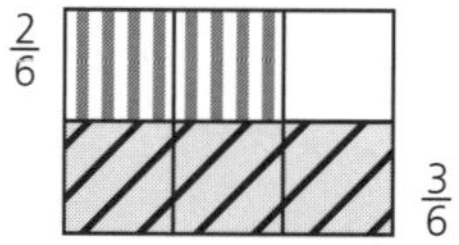

This looks like a total of $\frac{5}{6}$ …

… but this looks like a total of $\frac{5}{12}$.

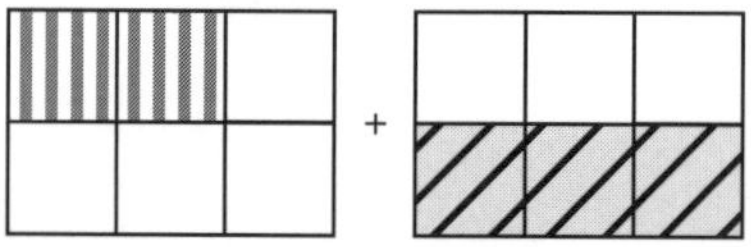

It is a common misconception to think the numerators and denominators can be added. Using two rectangles reinforces that misconception.

Ask your pupils to add other examples, such as $\frac{1}{8} + \frac{3}{8} = \frac{4}{8}$.

Ask them if they can see what $\frac{5}{11} + \frac{3}{11}$ is without drawing a diagram.

They should now be able to add any fractions with the same denominator.

Activity

Practise using folded rectangles of paper to add fractions with the same denominator.

3 Adding fractions with different denominators

Ask the pupils if they can add one banana + two oranges. The result is 1 banana and 2 oranges! The result of the addition cannot be expressed in an easy way. However, we can say that we have three fruits. We use the **common name**, fruit.

Similarly for fractions: $\frac{1}{3} + \frac{1}{5}$ means adding thirds and fifths.

To make the result easier to understand we need to add the same type of things. We need a **common denominator**.

First, consider fractions where one denominator is a multiple of the other.

For example, $\frac{1}{4} + \frac{3}{8}$.

Pupils cannot do this straightaway. They are trying to add eighths and quarters, which are different. Ask them, '*How can the denominators be made the same?*'
Answer: use equivalent fractions.

$\frac{1}{4} = \frac{2}{8} = \frac{3}{12} = \ldots$ One quarter equals two eighths.

So $\frac{1}{4} + \frac{3}{8} = \frac{2}{8} + \frac{3}{8} = \frac{5}{8}$.

Next, look at fractions where one denominator is not a multiple of the other.

For example, $\frac{1}{6} + \frac{3}{4}$.

Both fractions have to be changed so that they have the same denominator.

6 has multiples 6, 12, 18...
4 has multiples 4, 8, 12...

Notice 12 is a common multiple. The fractions can be expressed as twelfths.

So $\frac{1}{6} + \frac{3}{4} = \frac{2}{12} + \frac{9}{12} = \frac{11}{12}$.

For ease of arithmetic, it is always better to use the smallest, or lowest, common multiple (LCM). See Section B, Topic 18. Some people say that it is not necessary to find the LCM of the denominators in order to add (or subtract) fractions. 6×4 would also give a common denominator, but the arithmetic would involve larger numbers.

Activity

This sum can also be written as
$\frac{1}{6} + \frac{3}{4} = \frac{2+9}{12} = \frac{11}{12}$.
Which way do you think is easier for pupils to understand?

Activity

Work through this subtraction,
$\frac{7}{8} - \frac{5}{6}$.
Make 8×6 the denominator.
$\frac{7}{8} - \frac{5}{6} = \frac{42}{48} - \frac{40}{48} = \frac{2}{48} = \frac{1}{24}$.

This method means the fractions involve large numbers, and some division is required to simplify the result.

- Discuss this with other colleagues.
- Try the method out on some other examples.
- What do you think of it?
- Will your pupils find it simpler?

4 Subtracting fractions

A similar procedure is adopted with subtraction.
Start with fractions with same denominator.

three fifths – two fifths = one fifth

Show with the diagram:

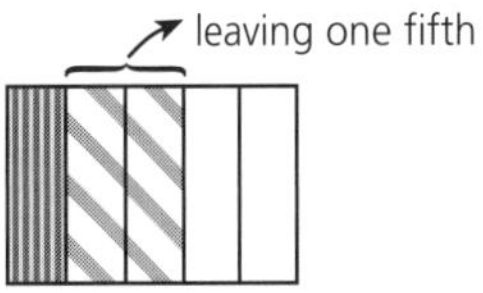

Topic 15 Multiplying and dividing fractions

1 Introduction

The method for multiplying and dividing fractions is simpler than the method for addition and subtraction.

(a) $\frac{a}{b} \times \frac{c}{d} = \frac{ac}{bd}$

Multiply the numerators together, multiply the denominators together.

(b) $\frac{a}{b} \div \frac{c}{d} = \frac{a}{b} \times \frac{d}{c}$

Turn the divisor upside down and multiply.

These rules should never be given to pupils without first explaining where they come from.

2 Fractions of quantities

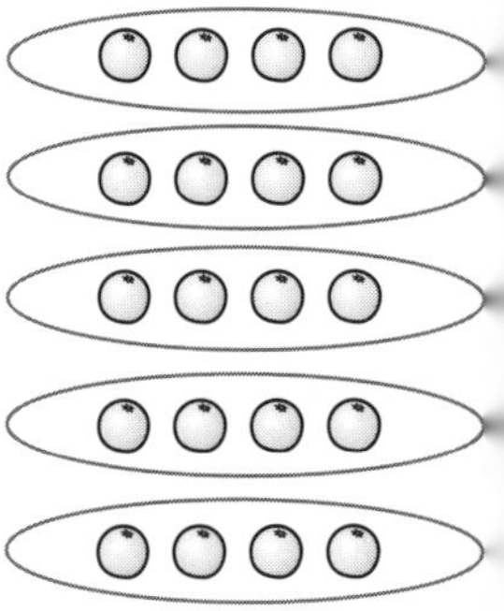

Start with examples of fractions with a numerator of 1.
Five girls share 20 oranges. Each girl gets one fifth of the oranges = four oranges.

So $\frac{1}{5}$ of $20 = \frac{1}{5} \times 20 = 4$

Three fifths of 20 oranges is $3 \times 4 = 12$ oranges.

So $\frac{3}{5}$ of $20 = \frac{3}{5} \times 20 = 3 \times \frac{1}{5} \times 20 = 3 \times 4 = 12$

In the same way,

$\frac{1}{6}$ of $42 = \frac{1}{6} \times 42 = 7$

and $\frac{5}{6}$ of $42 = \frac{5}{6} \times 42 = 5 \times \frac{1}{6} \times 42 = 5 \times 7 = 35$

Cancelling (dividing numerator and denominator by the same number — finding an equivalent fraction) reduces the size of the numbers in the calculation, and makes the calculation easier.

$\frac{5}{6}$ of $42 = \frac{5}{6} \times 42 = \frac{5}{\cancel{6}_1} \times \cancel{42}^{7} = 5 \times 7 = 35$

3 Multiplying fractions

The rule for multiplication can be explained by carefully chosen examples extending multiplication of whole numbers or by using diagrams.

3×8 means 3 lots of 8, or 3 times $8 = 24$

$3 \times \frac{2}{7}$ means 3 lots of $\frac{2}{7}$, or $3 \times$ two sevenths = six sevenths, $\frac{6}{7}$.

$$3 \times \frac{2}{7} = \frac{6}{7}$$

$\frac{1}{2} \times \frac{4}{7}$ means one half of four sevenths = two sevenths, $\frac{2}{7}$

$$\frac{1}{\cancel{2}_1} \times \frac{\cancel{4}^2}{7} = \frac{2}{7}$$

$\frac{1}{5} \times \frac{5}{6}$ means $\frac{1}{5}$ of 5 sixths = 1 sixth, $\frac{1}{6}$

$$\frac{1}{\cancel{5}_1} \times \frac{{}^1\cancel{5}}{6} = \frac{1}{6}$$

$\frac{3}{5} \times \frac{5}{6}$ means 3 times $\frac{1}{5}$ of $\frac{5}{6}$ = 3 times 1 sixth = $\frac{3}{6} = \frac{1}{2}$.

$$\frac{{}^1\cancel{3}}{\cancel{5}_1} \times \frac{{}^1\cancel{5}}{\cancel{6}_2} = \frac{1}{2}$$

$\frac{3}{5} \times \frac{1}{2}$ means 3 times $\frac{1}{5}$ of $\frac{1}{2}$ = 3 times $\frac{1}{10} = \frac{3}{10}$

$$\frac{3}{5} \times \frac{1}{2} = \frac{3}{10}$$

These examples demonstrate the rule, and use cancelling.

Using diagrams

For example: $\frac{2}{3}$ of $\frac{4}{5}$.

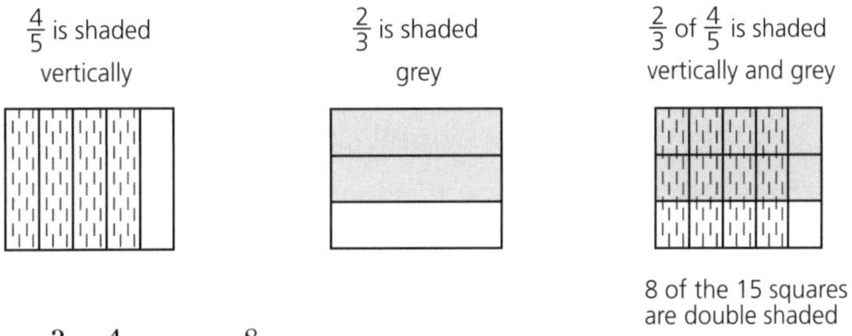

$\frac{2}{3} \times \frac{4}{5}$ means $\frac{8}{15}$

This can also be effectively demonstrated using translucent paper. Make two rectangles the same size. Shade the fractions in different directions. Place one rectangle exactly on top of the other one.

Some other examples:

$\frac{1}{2} \times \frac{1}{2} = \frac{1}{4}$ $\quad$ $\frac{3}{5} \times \frac{3}{4} = \frac{9}{20}$ $\quad$ $\frac{1}{3} \times 12 = 4$ $\quad$ $\frac{1}{3} \times \frac{1}{4} = \frac{1}{12}$

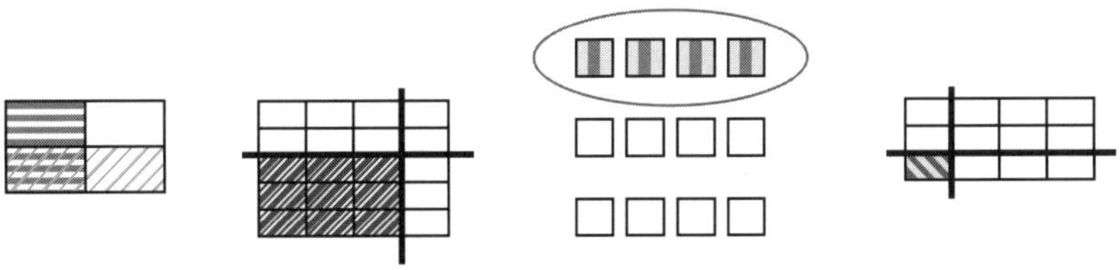

From these examples, and similar ones, the pupils can develop the rule.

So $\frac{5}{9} \times \frac{3}{10} = \frac{15}{90}$.

This can be simplified:

$\frac{15}{90} = \frac{5}{30} = \frac{1}{6}$,

by dividing the numerator and the denominator by 3 and then by 5.

The calculation is even simpler if we cancel (divide) before multiplying:

$$\frac{{}^{1}\cancel{5}}{\cancel{9}_{3}} \times \frac{{}^{1}\cancel{3}}{\cancel{10}_{2}} = \frac{1}{6}.$$

The method has been established, and justified.

4 Dividing by fractions

The rule '**When dividing by a fraction, turn it upside down and multiply**' was discussed in Section A Unit A1. It can also be justified by using well-chosen examples.

What is $10 \div 2$? In other words, how many 2s are there in 10?

Answer: 5

What is $3 \div \frac{1}{4}$? In other words, how many $\frac{1}{4}$s are there in 3?

Cut 3 cakes into quarters.

There are 4 quarters in 1 (cake), so $1 \div \frac{1}{4} = 4$

There are 12 quarters in 3 (cakes), so $3 \div \frac{1}{4} = 12$

However, $12 = 3$ lots of $4 = 3 \times 4 = 12$.

So $3 \div \frac{1}{4}$ is the same as $3 \times 4 = 12$.

This is easily understood for fractions with a numerator of 1. What about other fractions?

It is well known that $\frac{3}{4} \div \frac{3}{4} = 1$ because any number divided by itself is 1.

However, $\frac{3}{4} \times \frac{4}{3} = 1$ also, by cancelling, so dividing by $\frac{3}{4}$ has the same effect as multiplying by $\frac{4}{3}$.

So, for example, $\frac{2}{3} \div \frac{3}{4} = \frac{2}{3} \times \frac{4}{3} = \frac{8}{9}$

This first example was carefully chosen so that the answer is less than 1, and so that there is no cancelling.

Later examples can have cancelling and mixed number answers:

$$\frac{7}{8} \div \frac{5}{6} = \frac{7}{8} \times {}_{3}\frac{6}{\cancel{5}}$$

$$= \frac{{}^{4}\cancel{7}}{8} \times \frac{6}{5}$$

$$= \frac{21}{20}$$

$$= 1\frac{1}{20}$$

Activity

Construct a set of fraction division calculations that have:

- no cancelling and answers < 1
- cancelling and answers < 1
- answers > 1 and/or cancelling.

Topic 16 Adding and subtracting mixed numbers

1 Introduction

Fractions such as $\frac{13}{4}$ are sometimes called **vulgar** fractions or **top-heavy** fractions. In this book, they are just called fractions. Mixed numbers are numbers that are a combination of a whole number and a fractional part with a numerator smaller than its denominator. Pupils need to be taught that mixed numbers are written without the addition sign between the whole number and the fractional part. For example $8\frac{3}{4}$ means $8 + \frac{3}{4}$. The previous units carefully avoided fractions larger than one so that the basic ideas of the four operations could be discussed. However, mixed numbers will arise very soon — when adding fractions (such as $\frac{3}{4} + \frac{3}{4} = 1\frac{1}{2}$), when finding a fraction of a quantity (such as $\frac{3}{4}$ of $6 = 4\frac{1}{2}$) and when dividing a larger fraction by a smaller one (such as $\frac{3}{4} \div \frac{1}{2} = 1\frac{1}{2}$).

Pupils will find the last two examples puzzling. Their experience of whole numbers suggests to them that when you add or multiply numbers together, the result is a larger number. When you divide, the result is smaller. With fractions, this is no longer always true. Multiplying by a fraction, such as $\frac{1}{4} \times 6$ is another way of writing $6 \div 4$, so it should not be surprising that the answer is smaller. Whether a division results in a number greater than or smaller than 1 depends on the relative size of the two numbers. $10 \div 5$ is greater than 1 because $10 > 5$. $5 \div 10$ is smaller than 1 because $5 < 10$. In the same way, $\frac{3}{4} \div \frac{1}{2} > 1$ because $\frac{3}{4} > \frac{1}{2}$.

2 Changing fractions to mixed numbers

Start with fractions that can be written as whole numbers.

For example, $\frac{1}{4}$ of $12 = \frac{12}{4} = \frac{3}{1} = 3$ and $\frac{24}{6} = \frac{4}{1} = 4$

Then choose fractions (using the same denominators) that will not reduce to whole numbers.

So $\frac{13}{4} = \frac{12+1}{4} = \frac{12}{4} + \frac{1}{4} = 3 + \frac{1}{4} = 3\frac{1}{4}$

$\frac{29}{6} = \frac{24+5}{6} = \frac{24}{6} + \frac{5}{6} = 4 + \frac{5}{6} = 4\frac{5}{6}$

Compare this with division:

$13 \div 4 = 3$ remainder 1 (quarter) $29 \div 6 = 4$ remainder 5 (sixths)

From examples such as these, it can be seen that the way to change a top-heavy fraction into a mixed number is to divide the numerator by the denominator. The quotient is the whole number part, the remainder tells us the fractional part.

Thus $275 \div 7 = 39$ remainder 2, and $\frac{275}{7} = 39\frac{2}{7}$.

3 Changing mixed numbers to fractions

Use diagrams to do this. In this picture, three cakes

and half a cake are shaded. The number of cakes is $3\frac{1}{2}$.

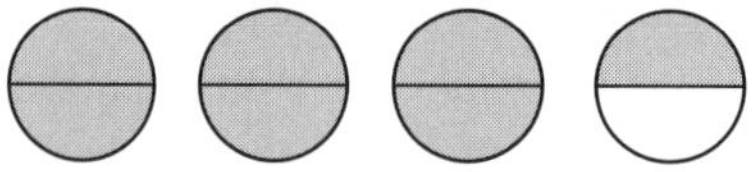

The three whole cakes have been divided into 6 halves.

$3\frac{1}{2}$ means $3 + \frac{1}{2}$ which is six halves plus one half $= \frac{6}{2} + \frac{1}{2} = \frac{7}{2}$

Similarly, $12\frac{3}{4}$ means $12 + \frac{3}{4} = \frac{48}{4} + \frac{3}{4} = \frac{51}{4}$.

In effect, we multiply the whole number part by the denominator of the fraction part, and then add the numerator of the fraction part. Multiplying can also be justified by noticing that when doing the opposite, changing fractions to mixed numbers, the opposite (dividing) was done.

4 Adding mixed numbers

As usual build up the complexity of the examples.

- Start with simple examples which have the same denominator and no 'carrying'.

whole numbers **eighths**

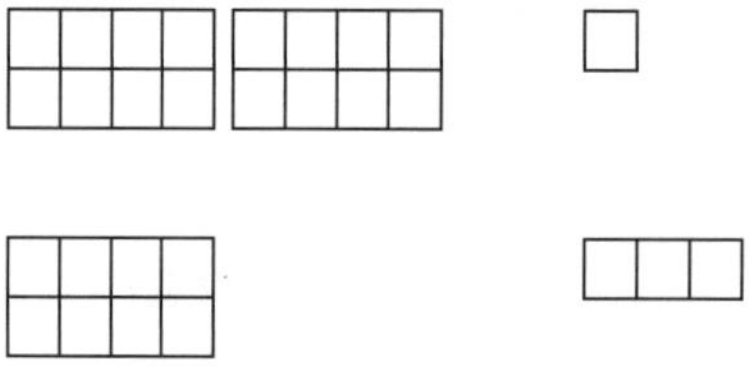

$$2\frac{1}{8} + 1\frac{3}{8} \text{ is clearly } (2 + 1) + \frac{1}{8} + \frac{3}{8} = 3\frac{4}{8} = 3\frac{1}{2}$$

- Then proceed to examples involving carrying. These can be done with diagrams if necessary:

  For example:

$$2\frac{7}{8} + 1\frac{3}{8} \text{ is clearly } (2 + 1) + \frac{7}{8} + \frac{3}{8} = 3\frac{10}{8} = 3 + 1\frac{2}{8} = 4\frac{1}{4}.$$

- Finally, include examples in which denominators differ.

5 Subtracting mixed numbers

Teaching subtraction follows a similar procedure. Start with examples with the same denominator and with the fraction of the number being subtracted smaller than the other fraction.

For example, $3\frac{5}{6} - 2\frac{1}{6}$.

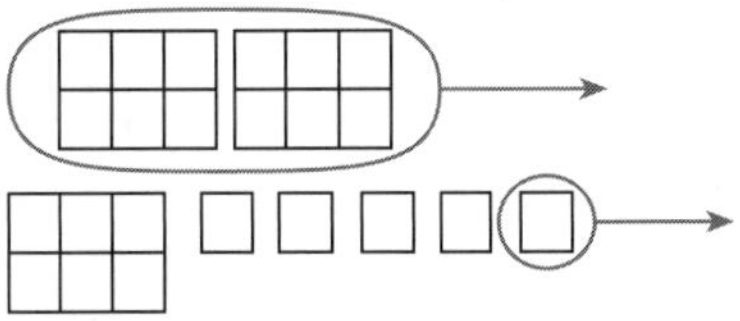

The picture shows $2\frac{1}{6}$ being taken away from $3\frac{5}{6}$, leaving $1\frac{4}{6} = 1\frac{2}{3}$.

The numbers and the fractional parts are worked out separately.

The difficulties, of course, arise when there is a need to 'borrow'. Once again, the first examples should have the same denominators.

There are two methods to choose from here.

(i) All the mixed numbers are changed to fractions:

$$3\frac{1}{8} - 1\frac{5}{8} = \frac{25}{8} - \frac{13}{8} = \frac{12}{8} = \frac{3}{2} = 1\frac{1}{2}$$

(ii) One of the whole numbers is rewritten as a fraction:

$$3\frac{1}{8} - 1\frac{5}{8} = (3 - 1) + \frac{1}{8} - \frac{5}{8} = 2\frac{1}{8} - \frac{5}{8}$$

$$= 1 + \frac{8}{8} + \frac{1}{8} - \frac{5}{8} = 1\frac{4}{8} = 1\frac{1}{2}.$$

The first method is simpler in this case, but it can lead to problems involving large numbers.

When the fractions have different denominators and the second fraction is larger, subtraction by the first method is more difficult.

(i) $8\frac{1}{8} - 6\frac{5}{9} = \frac{65}{8} - \frac{59}{9} = \frac{585}{72} - \frac{472}{72} = \frac{113}{72} = 1\frac{41}{72}$.

The second method solution has smaller numbers in it.

(ii) $8\frac{1}{8} - 6\frac{5}{9} = (8 - 6) + \frac{1}{8} - \frac{5}{9} = 2\frac{1}{8} - \frac{5}{9}$

$$= 2\frac{9}{72} - \frac{40}{72} = 1 + \frac{72}{72} + \frac{9}{72} - \frac{40}{72}$$

$$= 1\frac{41}{72}$$

Note: It is possible to combine the two methods, by subtracting the whole numbers and then changing what is left to fractions:

$$8\frac{1}{8} - 6\frac{5}{9} = 2\frac{1}{8} - \frac{5}{9} = \frac{17}{8} - \frac{5}{9} = \frac{153}{72} - \frac{40}{72} = \frac{113}{72} = 1\frac{41}{72}.$$

Activity

Consider which of the three methods is best to use.

Do this by considering some examples and discussing them with your colleagues.

Topic 17 Multiplying and dividing mixed numbers

1 Introduction

The methods for multiplying and dividing mixed numbers are simpler than those for adding and subtracting them. Nevertheless, pupils often get problems wrong because they muddle the different methods. They treat the whole number part and the fractional part separately when they should change all the numbers to fractions before multiplying. Teaching understanding rather than rote learning enables pupils to chose the correct methods.

2 Multiplying mixed numbers together

Pupils learnt, when adding mixed numbers, to add the whole number parts together and to add the fraction parts together. Often they transfer this rule to multiplying whole numbers.

For example, they may try: $2\frac{1}{2} \times 1\frac{3}{5} = (2 \times 1)(\frac{1}{2} \times \frac{3}{5}) = 2\frac{3}{10}$.

They need to be convinced that this method is wrong. For one thing, point out that

$2 \times 1\frac{3}{5} = 1\frac{3}{5} + 1\frac{3}{5} = 2\frac{6}{5} = 3\frac{1}{5}$. This is larger than $2\frac{3}{10}$.

Which do they think should be larger, $2\frac{1}{2} \times 1\frac{3}{5}$ or $2 \times 1\frac{3}{5}$?

Having convinced them that 'something is wrong', point out that in $2 \times 1\frac{3}{5}$, there are 2 ones and 2 three-fifths. The 2 multiplies both parts of the other number.

To illustrate 'a mixed number' × 'a mixed number', choose a pair of fractions that make drawing a diagram simple, e.g. $2\frac{1}{4} \times 3\frac{1}{2}$.

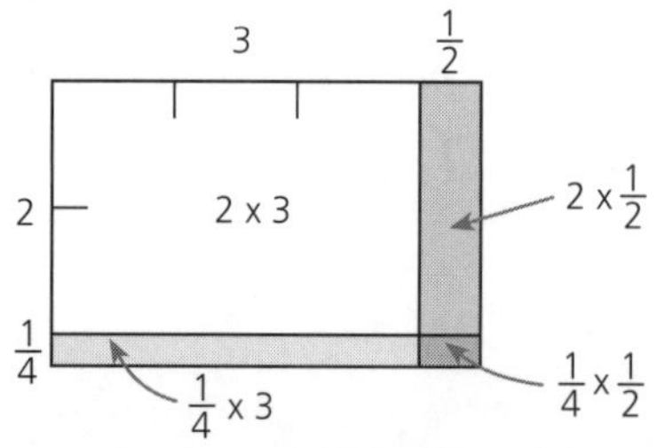

2 is multiplied with 3 and $\frac{1}{2}$.

$\frac{1}{4}$ is multiplied with 3 and $\frac{1}{2}$.

Working this out as $(2 \times 3) + (2 \times \frac{1}{2}) + (\frac{1}{4} \times 3) + (\frac{1}{4} \times \frac{1}{2})$ would be very tedious. It is much simpler to convert to fractions and multiply:

$$2\frac{1}{4} \times 3\frac{1}{2} = \frac{9}{4} \times \frac{7}{2} = \frac{63}{8} = 7\frac{7}{8}.$$

3 Dividing mixed numbers

The method of doing these calculations follows from the methods for dividing fractions and for multiplying mixed numbers.

For example, since $2\frac{1}{4} \times 3\frac{1}{2}$ is worked out by changing the mixed numbers to fractions and then multiplying, it is natural to work out $2\frac{1}{4} \div 3\frac{1}{2}$ by changing it to fractions, inverting and multiplying:

$$2\frac{1}{4} \div 3\frac{1}{2} = \frac{9}{4} \div \frac{7}{2} = \frac{9}{\cancel{4}_2} \times \frac{\cancel{2}^1}{7} = \frac{9}{14}$$

A common error is to divide the whole number and fractions separately.
For example, it is incorrect to write

$$12\frac{1}{4} \div 3\frac{1}{2} = (12 \div 3)\left(\frac{1}{4} \div \frac{1}{2}\right) = (4)\left(\frac{1}{\cancel{4}_2} \times \frac{\cancel{2}^1}{1}\right) = 4\frac{1}{2}.$$

This can be shown to be wrong by multiplying back again:

$$4\frac{1}{2} \times 3\frac{1}{2} = \frac{9}{2} \times \frac{7}{2} = \frac{63}{4} = 15\frac{3}{4}, \text{ not } 12\frac{1}{4}.$$

Topic 18 Multiples and factors

1 Introduction

The multiples and factors of numbers are useful tools in arithmetic and, later on, in algebra. (In this unit, the word *number* refers to the natural numbers 1, 2, 3, 4, 5, 6, 7, 8... .)

2 Multiples

Pupils can easily state the multiples of a small number by recalling the appropriate line in the multiplication table. The multiples of 3 are 3, 6, 9, 12, 15, 18, 21, 24, 27, ... The multiples of 2 are the **even** numbers 2, 4, 6, 8, 10, 12, 14, 16, 18, ... (The numbers which are not even are the **odd** numbers 1, 3, 5, 7, 9, 11, 13, 15, 17, 19, 21, ...) Pupils need to learn these important words. Even numbers can be shown as two equal groups of objects.

For example, 10 = $\begin{matrix} x\,x\,x\,x\,x \\ x\,x\,x\,x\,x \end{matrix}$.

Odd numbers cannot; there is always one (odd one) left over.

3 Tests for divisibility

It is easy to find out whether any number is a multiple of another by using a calculator. However, there are tests for divisibility, for deciding whether a number is a multiple of another number, which do not require a calculator. These tests are:

1 All natural numbers are multiples of 1.

2 Multiples of 2 end in 2, 4, 6, 8, or 0.

3 A number is a multiple of 3 when the sum of the digits is a multiple of 3.
For example, 1452. 1 + 4 + 5 + 2 = 12, which is a multiple of 3, so 1452 is a multiple of 3.
The sum of the digits of 1453 is 13, which is not a multiple of 3, so 1453 is not a multiple of 3.

4 The multiples of 4 are 4, 8, 12, 16, ... 100, 104, 108, ... 1000, 1004, 1008 ... A number is a multiple of 4 if the last two digits are a multiple of 4.
Thus 5672 is a multiple of 4 because 72 is a multiple of 4.

5 The multiples of 5 are 5, 10, 15, 20, 25, 30 .... Multiples of 5 end with a 5 or 0.

6 The multiples of 6 are 6, 12, 18, 24, 30, ... They are multiples of 2 **and** 3 so they end in 2, 4, 6, 8 or 0 **and** the sum of their digits is a multiple of 3.

7 The multiples of 7 are 7, 14, 21, 28, ... Although there is a rule to test if a number is a multiple of 7, it is complicated and does not help very much.

8 The multiples of 8 are 8, 16, 24, ... 96, 104, 112, ... 992, 1000, 1008, 1016, ... A number is a multiple of 8 when the last three

> ***Activity***
>
> Find out why the tests work.
>
> There is also a test for 11. Try to find out what it is.
>
> Make up tests for 15 and 18.
>
> Can you find a test for 7?

digits are a multiple of 8. Thus 5672 is a multiple of 8 because 672 is a multiple of 8.

9 A number is a multiple of 9 when the sum of the digits is a multiple of 9.
For example, 1458. 1 + 4 + 5 + 2 = 18, which is a multiple of 9, so 1458 is a multiple of 9.
The sum of the digits of 1459 is 19, which is not a multiple of 9, so 1459 is not a multiple of 9.

10 The multiples of 10 are 10, 20, 30, ... 100, 110, ... They all end in 0.

The pupils can tell the teacher the simple tests, for 2, 5 and 10. The other tests are not so obvious, but are easily learned.

4 Composites and primes

Multiples of other numbers are sometimes called **composite** numbers or **rectangular** numbers. 10 = 2 × 5 can be thought of as being composed of (made from) 2 lots of 5.
It can be shown as a rectangle:

x x x x x
x x x x x

Pupils can be asked to draw other composite numbers as rectangles. Some numbers can only be drawn as lines. For example 7 is x x x x x x x. These numbers are **prime numbers** or **primes**.

> **Note:**
> **A prime number** is a number whose only factors are 1 and itself. Pupils often think that 1 is a prime number but 1 is excluded from the list of primes for a very good reason.
>
> There is a rule which says that every composite number can be expressed as a product of primes in one way only. For example, 20 = 2 × 2 × 5 cannot be written as a product of prime numbers in any other way. If 1 were a prime number there would be an infinite number of ways of writing 20 as a product of primes, such as 1 × 2 × 2 × 5 and 1 × 1 × 2 × 2 × 5.
>
> A reason for excluding 1 that is easier for pupils to understand is that a prime number has exactly two factors, 1 and itself. 13, for example has factors 1 and 13. However, 1 has only one factor.

Pupils can find prime numbers by using the 'sieve of *Eratosthenes*'. They write out numbers in a table. Here it is done for numbers from 1 to 30. It is often done for numbers from 1 to 100.

Cross out 1.	~~1~~	2	3	~~4~~	5	~~6~~
Leave 2 but cross out the multiples of 2.	7	~~8~~	~~9~~	~~10~~	11	~~12~~
Leave 3 but cross out the multiples of 3.	13	~~14~~	~~15~~	~~16~~	17	~~18~~
Leave 5 but cross out the multiples of 5,	19	~~20~~	~~21~~	~~22~~	23	~~24~~
and so on.	~~25~~	~~26~~	~~27~~	~~28~~	29	~~30~~

The numbers that are left are the prime numbers 2, 3, 5, 7, 11, 13, 17, 19, ...
The sieve can be used to find larger primes.

A game for introducing primes

This game is for two players with 20 counters or bottle tops. Pupil A chooses a number of bottle tops and gives them to B. If pupil B can make a rectangle, B scores a point. If not, A scores a point. They take it in turns choosing a number. The same number cannot be used twice and the game ends when all the numbers from 2 to 20 have been used. Pupils will realise that the winning numbers are 2, 3, 5, 7, 11, 13, 17, 19. These are the prime numbers.

5 Least common multiples

The multiples of 2 and 3 contain some common numbers.

Multiples of 2: 2, 4, **6**, 8, 10, **12**, 14, 16, **18**, 20, 22, **24**, 26, ...
Multiples of 3: 3, **6**, 9, **12**, 15, **18**, 21, **24**, 27, 30, 33, 36, ...

The numbers occurring in both lists are 6, 12, 18, 24, ... These are the **common multiples** of 2 and 3. The **least common multiple (LCM)** is the smallest of these, 6. Note that all the other common multiples are multiples of the LCM.

Finding the LCM of two numbers

The simplest way for pupils to find the LCM of two numbers is to list the first few multiples of the numbers.

For example: 18 has multiples 18, 36, 54, 72, 90, 108, ...
24 has multiples 24, 48, 72, ...

There is no need to go any further as 72 is in the other list. The LCM is 72. It is quicker if the pupils write down only the first few terms of each list and then add terms to both lists until they find a common number.

Draw the attention of pupils to the useful fact that if one number is a multiple of another, then their LCM is the larger number.

For example,
120 is a multiple of 60, so the LCM of 60 and 120 is 120.

6 Factors

Some pupils confuse the words 'factors' and 'fractions', as they are very similar.

Multiples and factors are closely linked.
$24 = 6 \times 4$ so 24 is a *multiple* of 6, and 6 is a *factor* of 24.
If m is a multiple of n, then n is factor of m.

Note that 1 is a factor of every number and every number is a factor of itself.

Finding the factors of numbers

There are several methods. One way is to divide the number in turn by 2, 3, 4, 5 …

For example to find the factors of 48:

$48 \div \mathbf{1} = \mathbf{48}$
$48 \div \mathbf{2} = \mathbf{24}$
$48 \div \mathbf{3} = \mathbf{16}$
$48 \div \mathbf{4} = \mathbf{12}$ $48 \div 5$ is not a whole number
$48 \div \mathbf{6} = \mathbf{8}$ $48 \div 7$ is not a whole number
$48 \div 8 = 6$, but the numbers have started repeating, we already have 6 and 8.
You only need to continue up to $\sqrt{48}$ — which is between 6 and 7.
Therefore, the factors of 48 are 1, 2, 3, 4, 6, 8, 12, 16, 24 and 48.

Finding the *prime* factors of numbers

Divide the number in turn by 2, 3, 5 … only moving on to the next prime factor when you cannot divide by the smaller one any more.

For example to find the prime factors of 180:

$180 \div \mathbf{2} = 90$
$90 \div \mathbf{2} = 45$ 2 is not a factor of 45
$45 \div \mathbf{3} = 15$
$15 \div \mathbf{3} = 5$ 3 is not a factor of 5
$5 \div \mathbf{5} = 1$

This can be set out as a division:

$$\begin{array}{r|r} 2 & 180 \\ \hline 2 & 90 \\ \hline 3 & 45 \\ \hline 3 & 15 \\ \hline 5 & 5 \\ \hline & 1 \end{array}$$

$180 = 2^2 \times 3^2 \times 5$
The *prime factors* of 180 are 2, 3 and 5.

The prime factors of a number can also be found with a **factor tree**.

For example, 180

Find any two factors of 180, such as 10 and 18

Find two factors of 10 (2 and 5) and 18 (3 and 6)

6 has factors 2 and 3.

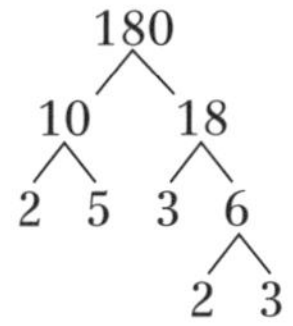

So $180 = 10 \times 18 = 2 \times 5 \times 3 \times 6 = 2 \times 5 \times 3 \times 2 \times 3 = 2^2 \times 3^2 \times 5$.

7 Highest common factor

The highest common factor (HCF) of two numbers is the largest factor that they both have.

For example,

78 has factors **1**, **2**, **3**, **6**, 13, 26, 39 and 78.
48 has factors **1**, **2**, **3**, 4, **6**, 8, 12, 16, 24 and 48.

The common factors of 78 and 48 are 1, 2, 3, and 6.
The HCF of 78 and 96 is 6.

Note that the HCF is a multiple of the other common factors.

Finding the HCF of two numbers

There are several ways that the pupils can do this. They can list all the factors of the two numbers, as in the example above. This is the easiest method to understand, but it can be quite lengthy.

Another way is to ask if there is any number that divides exactly into both numbers.

For example, what is the HCF of 144 and 108?
Can the pupils think of a number that divides exactly into both 144 and 108?
2 does (so 2 is a common factor), divide both numbers by 2:

$$\begin{array}{r|rr} 2 & 144 & 108 \\ \hline & 72 & 54 \end{array}$$

Is there a number that divides exactly into 54 and 72?
6 does (so 6 is another common factor), divide by 6:

$$\begin{array}{r|rr} 6 & 72 & 54 \\ \hline & 12 & 9 \end{array}$$

Is there a number that divides exactly into 12 and 9?
3 does (so 3 is another common factor), divide by 3:

$$\begin{array}{r|rr} 3 & 12 & 9 \\ \hline & 4 & 3 \end{array}$$

Is there a number that divides exactly into 4 and 3? No.
So, the HCF of 144 and 108 is $2 \times 6 \times 3 = 36$.
This can be written more briefly as

$$\begin{array}{r|rr} 2 & 144 & 108 \\ \hline 6 & 72 & 54 \\ \hline 3 & 12 & 9 \\ \hline & 4 & 3 \end{array}$$

Choosing different divisors does not change the answer:
12 divides into 144 and 108
3 divides into 12 and 9

$$\begin{array}{r|rr} 12 & 144 & 108 \\ \hline 3 & 12 & 9 \\ \hline & 4 & 3 \end{array}$$

> ***Activity***
>
> Find the LCM and HCF of 6 and 8.
>
> What is the result of the HCF × the LCM?
>
> Repeat for 10 and 15. What do you notice?
>
> Can you prove the result?

The HCF of 144 and 108 is $12 \times 3 = 36$, as above.

This approach can also be used to find the LCM of two numbers. It is product of the numbers down the left-hand side and along the foot of the calculation. The LCM of 144 and 108 is $2 \times 6 \times 3 \times 4 \times 3 = 432$.

However the reason why this gives the LCM is more difficult to understand than the methods quoted above, and because it is so similar to the method of finding the HCF, pupils often become confused.

Topic 19 Mental mathematics

1 Introduction

In Primary school, children learn the formal written algorithms for adding, subtracting, multiplying and dividing numbers. However, adults do not use these methods for most of the arithmetic they do. They use a calculator. They also calculate mentally, and teachers should help children to do this at school. For many calculations, mental methods are the simplest.

For example, 300 – 6 is more easily done mentally than by using the written algorithm.

Mental methods can be used to check answers found by other methods. Because they are usually methods the pupils understand (they can often develop the methods themselves), it is a good way for them to learn how numbers work.

Games have an important role in the teaching of maths. They encourage pupils to work mentally. They do not specify a method so pupils are not tied to a particular method. This gives pupils the opportunity to develop their own mental methods.

This unit looks at some ways in which pupils can learn to do calculations quickly, in their heads. The examples start with small numbers. Pupils can extend these methods to larger numbers. Mental maths is about developing a *range* of strategies to solve problems — the written algorithm is not the only way. Ian Thompson of Newcastle University has written extensively about mental calculation strategies.

Note that mental work is for *all* pupils. Encourage all pupils to have a go. Those pupils who have the most difficulty with the formal methods have the most to gain from different approaches. Mental methods can precede written methods. They can be used to encourage an active understanding of place value.

Do not reduce the teaching of this topic to the transmission of a set of rules. This would make it no different from ordinary computations and it may confuse pupils if they misapply them. Encourage pupils to develop their own methods.

2 Addition and subtraction

Pupils add and subtract mentally in various ways. The methods below are illustrated with two-digit numbers and extended to larger numbers.

Counting in ones

For example, 14 + 28 = 42.

- Pupils (especially young children) may count all the numbers when they have to find the total of 14 and 28 things or drawings.

- Sometimes they will 'count on' from the first or the larger number. This method is very effective when adding a small number to a larger number, such as 2486 + 7 = 2493. Counting on 7 from 2486 gives the answer.
- They can also 'bridge through 10' 2486 + 4 = 2490, then 3 more gives the answer.
- Pupils will come to realise that there is no need to count on laboriously. Knowledge of the 'number facts' will enable them to give answers much more quickly.
- To subtract, pupils can count back from the larger number. In order to calculate 38 – 6, they count back 6 places: 37, 36, 35, 34, 33, 32. This method can be used for 4000 – 6. Pupils will come to realise that since 10 – 6 = 4, then 4000 – 6 = 3994.

Compatible numbers

Numbers such as 7 and 3 which have a sum of 10, or 62 and 38, which have a sum of 100 are sometimes called **compatible.** Recognising compatible numbers can be used to facilitate addition.

For example, since 3 + 7 = 10, 2483 + 7 has a 0 in the Ones column, and the Tens column increases by 1.
Similarly, 362 + 538 = 100 + 300 + 500 = 900.

Complementary addition

Pupils can count on to solve subtraction problems, using **complementary addition** and the **partitions** of 10 (9 +1, 2 + 8, 3 + 7, 4 + 6, 5 + 5). The Dutch call this **adding to 10**.

- Start with numbers close together, such as 64 – 58.
  Add 2 to 58 to make 60, then add 4 more. 64 – 58 = 2 + 4 = 6.
- Then use examples where the numbers are further apart:
  For example,

170 – 106	Start from 106, add 4 to make 110, add 60 to make 170. So 170 – 106 = 4 + 60 = 64.
4567 – 4559	Count on 1 to 4560 and 7 to 4567. So 4567 – 4559 = 1 + 7 = 8.
6160 – 5972	5972 + 28 = 6000 and 6160 – 6000 = 160. So 6160 – 5972 = 28 + 160 = 188.

This method can be extended to decimals.

61.60 – 59.72 = 0.28 + 1.60 = 1.88.

(This method is similar to the method for working out the time from 9.30 a.m. to 3.15 p.m.)

Consider tens and ones separately

This is sometimes called **partitioning.**

For example, 24 + 48 = 72.
20 + 40 = 60, 8 + 4 = 12, so 24 + 48 = 60 + 12 = 72

This can be extended to larger numbers.

For example,

508 + 406   500 + 400 = 900, 8 + 6 = 14,
so 508 + 406 = 914

56 – 24 = 32   50 – 20 = 30, 6 – 4 = 2,
So 56 – 24 = 30 + 2 = 32

This method is more difficult when the subtraction requires regrouping, such as in 86 – 48. Children can avoid this by reasoning that 86 – 46 = 40 and there are still 2 more to take away. The answer is 40 – 2 = 38.

Counting in tens and ones

This is sometimes called **sequencing.**

For example, 24 + 48 = 72.
Add the tens then the ones.
24 + 40 = 64. 64 + 8 = 72.

This method uses the fact that adding a single digit to 2 digits, such as 64 + 8, is easier mentally. (Some pupils may add the ones and then the tens.)

Some pupils will start from 24, then jump in steps of 10: 34, 44, 54, 64, before adding the final 8.

Subtraction can be done by counting back or on.

For example, 54 – 28 = 26.
54 – 20 = 34, 34 – 8 = 26

Some pupils will start from 54 then count down in steps of 10: 54, 44, 34, before subtracting the final 8.

Some pupils will count on from 28: 28, 38, 48, then count on 6 to 54. The total counted on is 20 + 6 = 26.

A mixture of partitioning and sequencing is to add tens first, then one ones digit, then the other.

For example, 24 + 48.
20 + 40 = 60, 60 + 4 = 64, 64 + 8 = 72

This is sometime called the **split-jump method.** This method can be extended to larger numbers.

For example, 508 + 406.
500 + 400 = 900, 900 + 8 = 908, 908 + 6 = 914

Subtracting the tens first

38 – 14 can be worked out by taking 10 from 38 to make 28, then 4 more, giving 24.
Again the method can be extended to larger numbers, such as 588 – 482.

580 – 480 = 100, 8 – 2 = 6, so the answer is 106.

> ***Activity***
>
> Practise these different methods of mental addition and subtraction.
>
> Practise until you are able to work out these calculations quickly and accurately. Extend to decimals.

Balancing

These are also called **compensation methods**, and can be very effective. It helps with mental calculations to notice that 9 is one less than 10 (6 + 10 = 16, so 6 + 9 is one less, 15), 8 is 2 less than 10 so 7 + 8 is 2 less than 17.

Again this can be extended to larger numbers. For example,

29 + 52	29 is one less than 30. 30 + 52 = 62, so 29 + 52 = 81
499 + 187	500 + 187 = 687, so 499 + 187 is 1 less, 686
34 – 18	18 is 2 less than 20, so the answer will be 2 more than if subtracting 20. 34 – 20 is 14, so 34 – 18 = 16

3 Multiplication

Some multiplication computations can be done mentally.

- **Multiplication by 10, 100, ...**

  Pupils should be able to multiply by 10 and 100 mentally.

  For example, $234 \times 10 = 2340 \quad 45 \times 100 = 4500$

- **Multiplication by 4**

  One method is to double the number and then double again.

  For example, 54×4.
  $54 \times 2 = 108$, $108 \times 2 = 216$

- **Multiplication by 9, 19, 29, 39, ...**

  Nine is 1 less than 10. To multiply a number by 9, multiply by 10 and subtract the number.

  For example, $48 \times 9 = 48 \times (10 - 1)$
  $48 \times 10 = 480$, $480 - 48 = 432$

  This method can be extended to multiplying by any number that ends in a 9. For example,

59×24	$60 \times 24 = 1440$, 59×24 is 24 less than 1440, $1440 - 24 = 1416$
99×68	$100 \times 68 = 6800$, 99×68 is 68 less than 6800, $6800 - 68 = 6732$

- **Multiplication by 8, 18, 28, 38, ...**

  Eight is 2 less than 10. To multiply a number by 8, multiply by 10 and subtract twice the number.

  For example, $48 \times 8 = 48 \times (10 - 2)$
  $48 \times 10 = 480$, $480 - 48 \times 2 = 480 - 96 = 384$

- **Multiplication by 11**

  Eleven is 1 more than 10. To multiply a number by 11, multiply by 10 and add the number.

  For example, $432 \times 11 = 432 \times (10 + 1)$
  $432 \times 10 = 4320$, $4320 + 432 = 4752$

- **Multiplication by 5**
  Since 5 = 10 ÷ 2, to multiply by 5, multiply by 10 and then divide by 2.

  For example, 44 × 5 = 440 ÷ 2 = 220.
- **Multiplication by 25**
  Since 25 = 100 ÷ 4, to multiply by 25, multiply by 100 and then divide by 4.

  For example, 64 × 25 = 6400 ÷ 4 = 1600.
- **Multiplication by 50**
  Since 50 = 100 ÷ 2, to multiply by 50, multiply by 100 and then divide by 2.

  For example, 88 × 50 = 8800 ÷ 2 = 4400.
- **Multiplication by 1.5**
  Multiplying by 1.5 or $1\frac{1}{2}$ is the same as adding half the number to the original number.

  For example, 24 × 1.5 = 24 + $\frac{1}{2}$ of 24 = 24 + 12 = 36.

4 Division

Division is the inverse of multiplication and mental methods of division often use this fact.

- **Dividing by 5** is the same as dividing by 10 and doubling.

  For example, $140 \div 5 = \frac{140}{10} \times 2 = 14 \times 2 = 28$.
- **Dividing by 25** is the same as dividing by 100 and multiplying by 4.

  For example, $1600 \div 25 = \frac{1600}{100} \times 4 = 16 \times 4 = 64$.
- **Dividing by a fraction** is the same as multiplying by its inverse.

  For example, $100 \div \frac{1}{4} = 100 \times \frac{4}{1} = 400$.

> ***Activity***
>
> Practise mental methods of multiplying and dividing.
>
> Try to find other rules that you can use.

5 Using these methods with pupils

Give the pupils many opportunities to practise mental calculations. This can be done through a short 'warm-up' session at the start of each lesson, using Flash Cards or other games. Choose your examples in advance, and work out the answers in advance. Include examples involving decimals.

For example: 6004 – 5992 = 12
6.004 – 5.992 = 0.012

An activity to develop number sense

Write a number less than 50 on the board. Ask the class, '*What can you tell me about the number?*'

Answers could be: '*odd/even*', '*a multiple of …*', '*a factor of …*', '*a prime/ square number*', '*makes 100 if you add….*', '*half the number is ….*', '*double the number is ….*', '*one more/less is …*', and so on.

Topic 20 Integers: introduction, addition and subtraction

1 Introduction

Pupils have to learn how integers (sometimes called **directed numbers**) behave when the four operations are used with them. In this unit the raised symbol '$^-$' is used for 'negative' and '–' (the minus sign) is used for the operation 'take away'. Negative four, for example, is written, $^-4$, to distinguish it from – 4 which means 'minus 4'. The difference between the two ideas is similar to the difference between having an overdraft at a bank of \$4 ($^-$\$4) and spending \$4 (– \$4). Similarly, the raised symbol '$^+$' is used for 'positive' and '+' is used for the plus sign.

2 Introducing integers

There are many ways to introduce the pupils to the idea of negative numbers. One way is through the number line. It is not a good idea to use many different examples at the same time but all approaches need to stress that the numbers have a **zero** point and an idea of **direction**. They should all lead to a number line with positive and negative numbers on them:

Examples of negative numbers in real-life can help to illustrate the concept.

	Zero point	Directional words +/–
Temperature	0°C	hot/cold, warmer/cooler
Altitude		
(i) Water level	surface level	above/below, up/down, higher/lower
(ii) Height of pupils	mean height	taller/shorter
Money	no money	richer/poorer, credit/debit, receive/pay
Time	Independence year lunchtime	after/ before, earlier/later, after/ before, earlier/later

- **Temperature** This approach is common in cold countries. Pupils are familiar with negative temperatures, and the thermometer is a number line that pupils have experienced before. Negative temperatures are not so familiar to pupils in tropical countries, apart from in refrigerators and freezers.
- **Altitude** can be used in several ways to establish the idea of positive and negative numbers and a starting point. Two are considered here. One is using the surface of water and

considering things above and below its surface. The other is to consider the heights of some people, and consider how much they are above or below the middle or median height.

For example:
A fish swimming at a depth of 5 metres below the surface of a lake is at a depth of $^{-}5$ metres.

A boat has a flag at the top of a mast 10 m long. The flag is $^{+}10$ m above the water surface.

A group of pupils can be used. The median height can be found, or a 'standard height' can be quoted. This height can be taken as the starting point. If the standard height is 1 m 55 cm, then a taller pupil with a height of 1 m 60 cm has a height of $^{+}5$ cm. A shorter pupil with a height of 1 m 45 cm has a height of $^{-}10$ cm.

Activity

Develop lessons to introduce the directed number line through each of these models.

Try them out. Decide which approach you like best.

- **Money** is a common example. Having no money is the zero point. Owing money is negative, and being in credit is positive.
- **Time** Any time can be taken as zero time. Independence year is a good choice. Three years before independence is $^{-}3$ years, 5 years after is $^{+}5$ years. A more humorous example is found by saying lunchtime is zero time. 2 hours before lunch is $^{-}2$ hrs.
  The Christian and Islamic calendars cannot be used, as in each case, the year before year 1 was not year 0, it was $^{-}1$.

3 Order of integers

Pupils are often puzzled by statements like $^{-}25 < {}^{-}20$. This statement can be interpreted as '$^{-}25$°C is colder than $^{-}20$°C', or 'a debt of \$25 means you are poorer than a person with a debt of \$20', and so on. On the number line, the rule is simple. A number is less than any number to the right of it. For example, $1 < 3$.

4 Addition

Addition should be taught by considering what is happening to the numbers on the number line. When pupils have understood this they can be shown a quick way to arrive at solutions. For example, a quick way to work out $(^{-}108) + (^{+}110)$ is to look at the numbers. The difference between 108 and 110 is 2. The sign of the result is positive, because the larger number 110 is positive,
so $(^{-}108) + (^{+}110) = {}^{+}2$.
Similarly, $(^{+}108) + (^{-}110) = {}^{-}2$, because 110 is negative.

Using the number line

The pupils will know that adding a positive number is a move to the right. The first step is to show that adding a negative number is a move to the left. One method is to show this on the number line.

- **Negative number plus positive number:** e.g. $^{-}4 + {}^{+}5$
  $^{-}4$ means 4 steps to the left and $^{+}5$ means 5 steps to the right. The result is 1 step to the right ($^{+}1$).

On the number line:

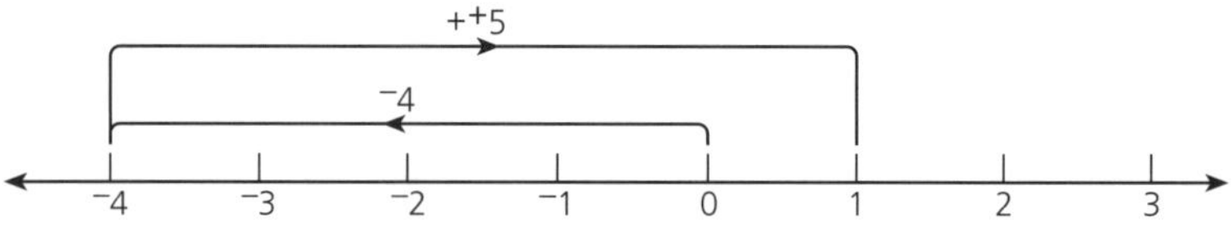

Note that the start is from 0 and not from $^{-}4$. By doing the problem in this way, it becomes a move to the left followed by a move to the right.

- **Positive number plus negative number:** e.g. $^{+}5 + {}^{-}4$
  Turning the sum around, $(^{+}5) + (^{-}4)$ will still have the same result:

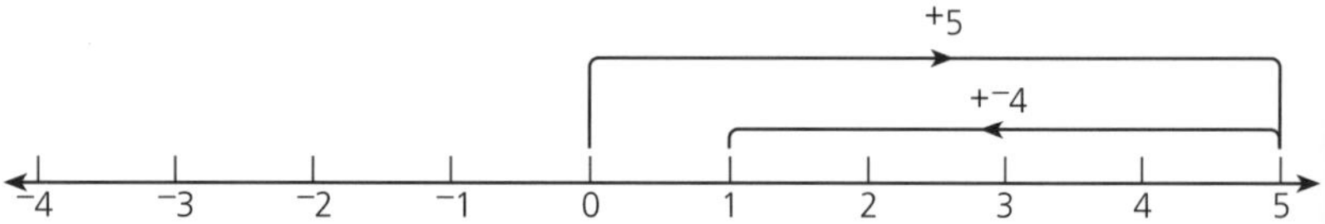

Start at zero, move 5 steps to the right. Then add $(^{-}4)$, which is 4 steps to the left.

Using cards

Another way to show the process is by using small cards, some labelled $^{+}1$, some labelled $^{-}1$. To make them more attractive, the positive and negative cards can be different colours.

The sum of one $^{+}1$ card and one $^{-}1$ is 0, because $^{+}1 + {}^{-}1 = 0$. (This can be shown on a number line).

The number $^{+}5$ is shown by five $^{+}1$ cards, and $^{-}4$ needs four $^{-}1$ cards. $^{+}5 + {}^{-}4 = {}^{+}1$ is shown thus:

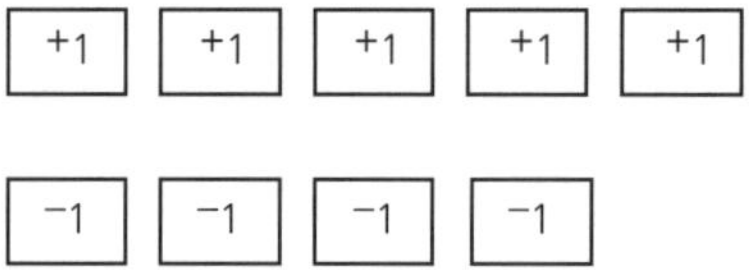

Four $^{-}1$ cards 'cancel out' four $^{+}1$ cards leaving one $^{+}1$ card.

Counters could be used instead of cards. Positive counters '+' would be one colour, and negative counters '–' would be another colour.

Inverses

Include examples in which integers are added to their inverses.

For example, $^{-}12 + {}^{+}12 = 0$.

This is important for understanding the next section, on subtracting integers.

5 Subtraction

Using cards, $^{-}8 - {}^{-}5 = {}^{-}3$ is easily shown: start with eight $^{-}1$ cards and take away five – 1 cards. This leaves three $^{-}1$ cards. But experience suggests that both the cards and the number line confuse many pupils trying to understand how to subtract negative numbers from positive numbers or positive numbers from negative numbers.

Activity

- Use the number line to calculate the following.
  (i) $^-4 + 2$ (ii) $^-2 - {}^+4$
  (iii) $^-8 + {}^+6$ (iv) $^+6 - {}^+4$
- Make small cards and label them $^+1$ and $^-1$. Use them to work out the following.
  (i) $^-4 + {}^+6$ (ii) $^-2 + {}^+10$
  (iii) $^-7 + {}^+4$

Repeat this with counters instead of cards.

The simplest way to teach the subtraction of integers is to show that subtracting an integer has the same effect as adding its inverse.

$^-4 - \mathbf{{}^-4} = 0$ because any number minus itself = 0

But $^-4 + \mathbf{{}^+4} = 0$ also, from the work on addition.

So $- {}^-4$ has the same effect as $+ {}^-4$.

Similarly $^-48 - {}^-48 = 0$ and $^-48 + {}^+48 = 0$ so $- {}^-48$ has the same effect as $+ {}^+48$.

A few examples like this and the rule **subtracting an integer is the same as adding its inverse** is established.

Topic 21 Integers: multiplication and division

1 Introduction

The rules for the multiplication and division of integers are much simpler than the rules for addition and subtraction. They can be shown in a simple table:

	Positive	**Negative**
Positive	Positive	Negative
Negative	Negative	Positive

2 Multiplication

The rules for multiplication can be developed in several ways.

$^{+}3 \times {}^{+}5$ is the same as $3 \times 5 = 15$.

Clearly, **positive × positive = positive**

$^{+}3 \times {}^{-}5$ can be thought of as 3 lots of $^{-}5$
and 3 times $^{-}5 = {}^{-}5 + {}^{-}5 + {}^{-}5 = {}^{-}15$

so, **positive × negative = negative**

This can be turned round and become negative × positive. We know multiplication is commutative (order doesn't matter), so:

$^{-}5 \times {}^{+}3 = {}^{+}3 \times {}^{-}5 = {}^{-}15$ i.e. **negative × positive = negative**

A few examples like these demonstrate that, always,

negative × positive = positive × negative = negative.

Negative × negative = positive is more difficult to explain. Some people justify the rule using sentences with a double negative in them. The sentence 'The ball is not not red' is taken as meaning 'The ball is red.' The two negatives cancel each other out. This is not entirely convincing, though it is effective as an aid to remembering the rule. Negation and 'negativeness' are not exactly the same. Some people use the example of friends and enemies: 'The enemy of my enemy is my friend.' is not necessarily true, and it is possible for a person to have two friends who hate each other!

A model for multiplying two negative numbers

Consider an item that becomes less valuable as time passes. For example, suppose a computer loses $20 in value each year.
Then, in 3 years time ($^{+}3$ years), the computer will be worth $60 less than it is worth now.
This can be expressed as $^{+}3 \times {}^{-}20 = {}^{-}60$.
Three years ago ($^{-}3$ years), however, the computer was worth $60 more than it is now.

This can be written as $^{-}3 \times {}^{-}20 = {}^{+}60$. It is not difficult to see that negative multiplied by negative will always be positive, since an item that is losing money will always have been worth more in the past.

Another approach is to consider the multiplication table for numbers from $^{-}3$ to $^{+}3$. Three-quarters of the table can be filled in using what has already been learnt about negative numbers. The only part of the table that has not been completed is the negative × negative part.

×	$^{-}3$	$^{-}2$	$^{-}1$	0	$^{+}1$	$^{+}2$	$^{+}3$
$^{+}3$	$^{-}9$	$^{-}6$	$^{-}3$	0	$^{+}3$	$^{+}6$	$^{+}9$
$^{+}2$	$^{-}6$	$^{-}4$	$^{-}2$	0	$^{+}2$	$^{+}4$	$^{+}6$
$^{+}1$	$^{-}3$	$^{-}2$	$^{-}1$	0	$^{+}1$	$^{+}2$	$^{+}3$
0	0	0	0	0	0	0	0
$^{-}1$				0	$^{-}1$	$^{-}2$	$^{-}3$
$^{-}2$				0	$^{-}2$	$^{-}4$	$^{-}6$
$^{-}3$				0	$^{-}3$	$^{-}6$	$^{-}9$

Look at the patterns of numbers in any of the lines. The left-hand vertical line contains the numbers $^{-}9$, $^{-}6$, $^{-}3$, 0. The next 3 numbers in this pattern should be $^{+}3$, $^{+}6$, $^{+}9$. By this argument, the rest of the numbers can be added to the table.
In every case, negative × negative is positive.

×	$^{-}3$	$^{-}2$	$^{-}1$	0	$^{+}1$	$^{+}2$	$^{+}3$
$^{+}3$	$^{-}9$	$^{-}6$	$^{-}3$	0	$^{+}3$	$^{+}6$	$^{+}9$
$^{+}2$	$^{-}6$	$^{-}4$	$^{-}2$	0	$^{+}2$	$^{+}4$	$^{+}6$
$^{+}1$	$^{-}3$	$^{-}2$	$^{-}1$	0	$^{+}1$	$^{+}2$	$^{+}3$
0	0	0	0	0	0	0	0
$^{-}1$	**$^{+}3$**	**$^{+}2$**	**$^{+}1$**	0	$^{-}1$	$^{-}2$	$^{-}3$
$^{-}2$	**$^{+}6$**	**$^{+}4$**	**$^{+}2$**	0	$^{-}2$	$^{-}4$	$^{-}6$
$^{-}3$	**$^{+}9$**	**$^{+}6$**	**$^{+}3$**	0	$^{-}3$	$^{-}6$	$^{-}9$

The weakness of this approach is that it is not a *justification* of the rule but merely a suggestion of a solution. Without understanding the reasons for the rule, some pupils misapply the rule and think that $^{-}3 + {}^{-}5 = {}^{+}8$ (using 'a negative and a negative make a positive'). Understanding rules should help pupils to apply them.

Activity

Consider the different approaches to the justification of the rules.

Decide which would be the best to use with your classes.

4 Division

Division is the inverse of multiplication and the rules for dividing integers can be arrived at in a simple way from the multiplication rules.

$3 \times 2 = 6$ can be expressed as a division as
$6 \div 3 = 2$ or as $6 \div 2 = 3$.

In the same way, $^{+}3 \times {}^{-}2 = {}^{-}6$ (positive × negative = negative) can be expressed as

$^{-}6 \div {}^{+}3 = {}^{-}2$ (negative ÷ positive = negative)
or as $^{-}6 \div {}^{-}2 = {}^{+}3$ (negative ÷ negative = positive).

Similarly, $^{-}3 \times {}^{-}2 = {}^{+}6$ (negative × negative = positive) can be expressed as

$^{+}6 \div {}^{-}3 = {}^{-}2$ (positive ÷ negative = negative).

The division rules are, therefore:

negative ÷ positive = positive ÷ negative = negative
positive ÷ positive = negative ÷ negative = positive

Topic 22 Introducing decimals

1 Introduction

Decimal fractions are an extension of the place value number system — hundreds, tens, ones, tenths, hundredths... They have become an important way of representing numbers for several reasons:

- Ordinary fractions can be quite difficult to work with. The rules of operations are not at all obvious and mixed numbers cause additional difficulties. Calculations with decimals are much easier, as they follow the same rules as calculations with whole numbers. All that is different is the inclusion of a decimal point and, sometimes, some zeros.
- Decimals and the metric system are closely connected. 5.4 cm is 5 cm 4 mm, for example.
- Calculations using calculators and computers are easier with decimal fractions than with other fractions.
- Money everywhere uses decimals. $1 = 100 cents, so 5 dollars 43 cents, for example, is written as the decimal $5.43.

However, decimals are another way of writing fractions and the children must understand that 0.34, for example, means $\frac{34}{100}$, if they are to carry out numerical calculations.

2 Conceptual difficulties with decimals

Many children do not realise that the number after the decimal point is less than one. This error is reinforced if they say 0.29, for example, as 'zero point twenty-nine.' Many pupils think that 0.75, for example, is greater than 0.8 and 4.90 is more than 4.9. The longer number is thought to be larger (which, of course, it is for whole numbers).

3 Introducing decimals

There are several ways to introduce decimals.

- **Money** Pupils will be familiar with counting money, buying and selling. In many countries, however, the value of the currency is so low that the division of the basic unit into hundredths is never used in practice. For those countries where it is possible, pupils will know what $14.63 means (for example, using dollars and cents). They will understand that 10 cents is $\frac{1}{10}$ of a dollar and 1 cent is $\frac{1}{100}$ of a dollar. $14.53 in place-value columns is:

Tens	Ones		tenths	hundredths
1	4	.	5	3

or $14.53 = 14 + \frac{5}{10} + \frac{3}{100}$

A difficulty with using money is that in money there are always exactly two decimal places. Fourteen dollars and fifty cents is written $14.50.

- **Place value** The decimal system can be explained by reminding pupils that in our counting system, 1243 means 1 thousand, 2 hundreds, 4 tens and 3 ones. The position of the number is important.

Thousands	Hundreds	Tens	Ones	
1	2	4	3	one thousand, two hundred and forty-three
5	0	0	0	five thousand
3	0	2	0	three thousand and twenty

Zeros are put in to show where a place has not been used.
Each position is one tenth of the position to its left.

$1000 \div 10 = 100$, $100 \div 10 = 10$, $10 \div 10 = 1$.

If this were continued to the right from the ones column, the next column would be tenths, $1 \div 10 = \frac{1}{10}$, one tenth. The next column would be hundredths, because $\frac{1}{10} \div 10 = \frac{1}{100}$, one hundredth, and so on.

Ones		tenths	hundredths	thousandths	
4	•	8			$4.8 = 4 + \frac{8}{10}$
4	•	8	5		$4.85 = 4 + \frac{8}{10} + \frac{5}{100} = 4\frac{85}{100}$
4	•	8	0	5	$4.805 = 4 + \frac{8}{10} + \frac{5}{100} + \frac{5}{1000} = 4\frac{85}{1000}$

The **decimal point** marks the place where numbers change to less than 1. Numbers to the **left** of the decimal point are **whole numbers**. Numbers to the **right** are **decimal fractions**.

- **Zeros** A zero contained *within* a decimal number, is important (or significant) — it tells us the positions of the other digits in the number.

For example, 24.03 is a **decimal number**.
24 is a whole number.
0.03 is a **decimal fraction** (often we just say decimal).

The zero tells us there are no tenths and the 3 is in the hundredths position.

24.630 is also a decimal number.
24 is a whole number.
.630 is a decimal fraction.

The zero gives us no more information about the position of the other numbers, but it does tell us there are no thousandths.

Other ways of representing decimals

2.7

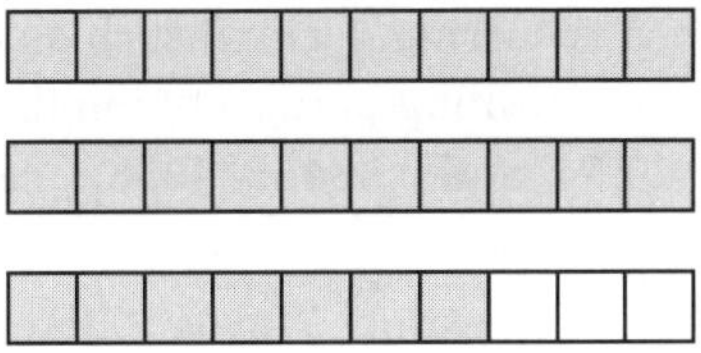

97.4

T 0 t

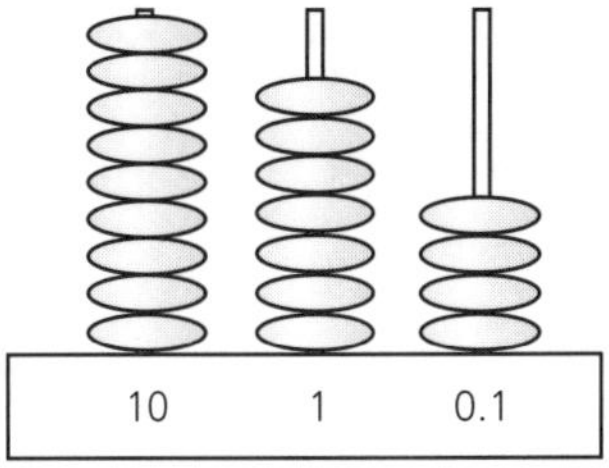

4 Changing a decimal to a fraction

This is quite easy to explain.

A number with 1 decimal place is a fraction with denominator 10.

For example: $0.4 = \frac{4}{100}$

A number with 2 decimal places is a fraction with denominator 100.

For example: $0.36 = \frac{36}{100}$ $\quad 0.04 = \frac{4}{100}$

(Pupils need to see examples with a zero in the tenths column.)

A number with 3 decimal places is a fraction with denominator 1000.

For example: $0.384 = \frac{384}{1000}$ $\quad 0.007 = \frac{7}{1000}$

Pupils will see the pattern that the number of zeros is the same as the number of places of decimals.

For example: $0.483\,216 = \frac{483\,216}{1\,000\,000}$ (six places)

5 Changing a fraction to a decimal

Pupils need to learn about two ways of doing this.

- In some cases, they can find an equivalent fraction with a denominator that is a power of ten:

  $\frac{4}{5} = \frac{8}{10} = 0.8 \qquad \frac{3}{25} = \frac{12}{100} = 0.12$

- In all cases, division can be used:

  $\frac{3}{8} \rightarrow 8\overline{)3.000}$ with quotient 0.375, so $\frac{3}{8} = 0.375$

6 Recurring decimals

In the example for $\frac{3}{8}$ above, the division ended at three decimal places, there was no remainder to continue the division. Other examples like this are $\frac{5}{8} = 0.625$ and $\frac{17}{40} = 0.425$.

Sometimes, however, the division goes on forever.

For example, $\frac{5}{6} = 0.8333333....,$

$\frac{6}{7} = 0.8571....$

Decimals that go on forever are called **non-terminating** decimals. Decimals that go on forever and have a repeating pattern to the numbers are called **recurring** decimals.

A recurring decimal such as 0.33333... can be written $0.\dot{3}$. One with a repeating pattern of numbers such as 0.5454.... can be written $0.\dot{5}\dot{4}$, and 0.3846384638... can be written $0.\dot{3}84\dot{6}$.

> **A question** Why is it that when a fraction is changed into a decimal, the decimal either stops or recurs?

7 Putting decimals in order

With whole numbers, the longer the number, the larger it is. For example, 1243 > 98. However, this is not true for decimals. 0.1234 < 0.98 but pupils often do not realise this. If they write the decimals as fractions with the same denominator, the idea is clearly explained:

$$0.1234 = \frac{1234}{10\,000},$$

$$0.98 = \frac{98}{100} = \frac{9800}{10\,000}.$$

0.98 is clearly larger.

There is a simpler way that pupils can use to order decimals. Look at the first decimal place. The value of first decimal place digit is bigger than all the other decimal places put together. The larger number has the larger digit in the first decimal place. Here 9 >1 so 0.98 > 0.1234.

If the digits in the first decimal place are the same, look at the second decimal place. So 4.11 > 4.1098. If the digits in second decimal places are also the same, look at the third decimal place, and so on.

Remind the pupils to compare the whole number first.

24.93 is not greater than 142.15!

Topic 23 Operations on decimals

1 Introduction

This topic looks at the teaching of addition, subtraction, multiplication and division of decimals. Addition and subtraction go together and can be taught at the same time. Multiplication and division are a little more complicated.

2 Addition and subtraction

The common mistake that pupils make is not to align the decimal points.

For example, 3.5 + 14 $\quad \begin{array}{r} 3.5 \\ +\ 1\,4 \\ \hline 4.9 \\ \hline \end{array} \qquad$ 8.63 – 4.2 $\quad \begin{array}{r} 8.6\,3 \\ -\ \ 4.2 \\ \hline 8\,2.1 \\ \hline \end{array}$

The pupils should be asked, '*Are these sensible answers?*'
What *are* sensible answers?

3.5 + 14 is a bit more than 3 + 14, or 17, so 4.9 cannot be correct.
8.63 – 4.2 is about 8 – 4, or 4, so 82.'something' cannot be correct.

Pupils can begin by adding simple decimals by changing them to fractions first, finding the answer and seeing a simpler way to proceed:

$$2.4 + 4.9 = 2\tfrac{4}{10} + 4\tfrac{9}{10} = 6\tfrac{13}{10} = 7\tfrac{3}{10} = 7.3$$

$$7.8 - 4.3 = 7\tfrac{8}{10} - 4\tfrac{3}{10} = 3\tfrac{5}{10} = 3.5$$

This can also be done with base-ten materials, or diagrams on paper.

2.4 + 4.9 is :

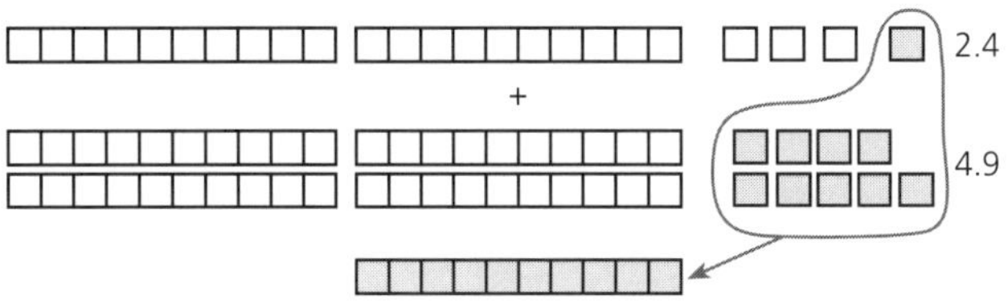

In early examples, avoid decimals with zeros in them. Also, the first subtractions should not involve decomposition.

Compare the examples above with 24 + 49 = 73 and 78 – 43 = 35.

Write them vertically:

$$\begin{array}{r} 2.4 \\ +\ 4.9 \\ \hline 7.3 \\ \hline \end{array} \qquad \begin{array}{r} 2\,4 \\ +\ 4\,9 \\ \hline 7\,3 \\ \hline \end{array} \qquad \begin{array}{r} 7.8 \\ -\ 4.3 \\ \hline 3.5 \\ \hline \end{array} \qquad \begin{array}{r} 7\,8 \\ -\ 4\,3 \\ \hline 3\,5 \\ \hline \end{array}$$

The pupils should be able to tell the teacher that adding or subtracting decimals is the same as adding or subtracting whole numbers: the tenths have to be in line, the hundredths have to be in line, and so on. The simplest way to ensure this is to put the decimal points in line.

3 Multiplication of decimals

The rule the pupils need to understand is:

Count the number of digits after the decimal point in the numbers being multiplied together. This is the number of digits after the decimal point in the answer.

For example, $24 \times 38 = 1012$, so $2.4 \times 3.8 = 10.12$
(two numbers after the decimal point).

- **Start by teaching pupils to multiply decimals by 10, 100, 1000.**

Pupils know that $28 \times 10 = 280$, $28 \times 100 = 2800$ …. Digits move one place to the left when multiplied by 10, two places when multiplied by 100… .
The same is true for decimals.

For example:

$$0.2 \times 10 = \frac{2}{10} \times 10 = 2$$

or 2 tenths $\times$ 10 = 20 tenths = 2

$$0.08 \times 10 = \frac{8}{100} \times 10 = \frac{8}{10}$$

or 8 hundredths $\times$ 10 = 8 tenths = 0.8.

Illustrate this by shading diagrams.

For example, 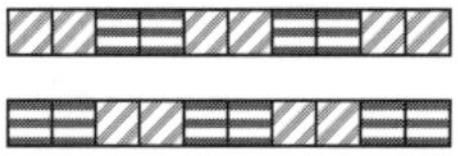 each shaded part is 0.2. 10 of them make 2.

Establish the rule in the same way for examples of multiplying by 100, 1000 … .

- **Multiply a decimal by a single whole number.**

Start with an example of multiplication which needs no 'carrying'.

For example:
1.2×3 $\quad 1.2 \times 3 = 1\frac{2}{10} \times 3 = 3\frac{6}{10} = 3.6$

Compare with $12 \times 3 = 36$.

$$\begin{array}{r} 1.2 \\ \times\ 3 \\ \hline 3.6 \end{array} \qquad \begin{array}{r} 12 \\ \times\ 3 \\ \hline 36 \end{array}$$

The procedure is the same, with a decimal point added.

Continue with examples involving carrying, such as 2.4×8 and 0.6×4. These can also be demonstrated using strips.

For example, $0.6 \times 4 = 2.4$

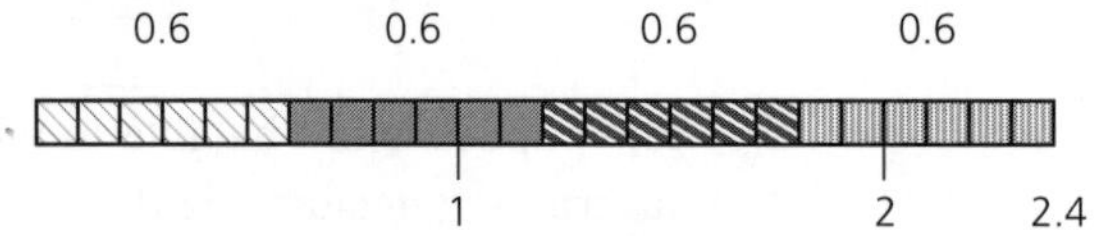

- **Multiply a decimal by a 2-digit whole number.**

  For example, 2.3×36

	T O t
6×3 tenths $(6 \times .3) = 18$ tenths =	1.8
6×2 ones $(6 \times 2) = 12$ ones =	1 2
30×3 tenths $(30 \times .3) = 90$ tenths =	9
30×2 ones $(30 \times 2) = 60$ ones =	6 0
	8 2.8

More briefly:

	2.3
	× 36
6×2.3	13.8
30×2.3	69.0
	82.8

Compare this with 23×36 by long multiplication:

$$\begin{array}{r} 23 \\ \times\, 36 \\ \hline 138 \\ 690 \\ \hline 828 \end{array}$$

The procedure is the same as for whole numbers, except for the added decimal point.

- **Multiply any decimal × any decimal.**

  Start with a few examples using fractions.

  For example: $5.2 \times 1.6 = \frac{52}{10} \times \frac{16}{10} = \frac{52 \times 16}{10} = \frac{832}{100} = 8.32$

  $$0.3 \times 0.05 = \frac{3}{10} \times \frac{5}{100} = \frac{15}{1000} = 0.015$$

  Compare with $52 \times 16 = 832$ and $3 \times 5 = 15$.

The pupils can see the computation is the same. Tell them to count how many digits there are after the decimal point in the example questions. There are the same number of digits after the decimal point in the products (or the answers). Why is this?

Simple examples such as $0.3 \times 0.5 = 0.15$ can be shown on a diagram:

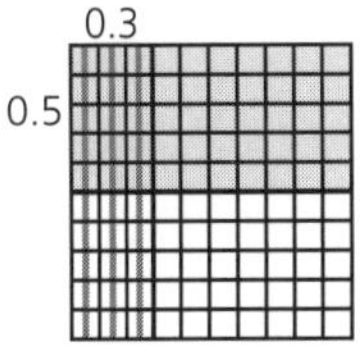

In early examples, avoid having 5 in the last decimal place multiplied by an even number.
For example, $5.5 \times 16 = 88$ appears to break the rule.
Once pupils have learnt what to do they will work out $5.5 \times 16 = 88.0$.

> **Discussion** Discuss why the multiplication rule works.

4 Division of decimals

- **Divide a decimal by a whole number.**

The procedure is the same as with ordinary division, but now there is a decimal point in the numbers. Start with simple examples, such as 4.35 ÷ 3.

Note that 435 ÷ 3 = 145. $3\overline{)435}$ = 145

Using fractions, $4.35 \div 3 = 4\frac{35}{100} \div 3 = \frac{435}{100} \times \frac{1}{3} = \frac{145}{100} = 1.45$

This is a very complicated way of doing it! 4.35 divided by 3 is the same calculation as 435 ÷ 3, but now there is a decimal point to think about:

$$\begin{array}{r} 1.45 \\ 3\overline{)4.35} \end{array}$$

- **Divide by a decimal.**

It is difficult to divide by a decimal using long division. For example, what is 48 ÷ 1.6? How many times does 1.6 go into 40? It is hard to work it out. So, try to change the calculation to one involving division by a whole number. To change 1.6 to a whole number, multiply by 10. To keep the value the same, multiply the numerator by 10 as well.

So $48 \div 1.6 = (48 \times 10) \div (1.6 \times 10) = 480 \div 16 = 30$.

Justify this with a simpler example:
$12 \div 3 = (12 \times 2) \div (3 \times 2) = 24 \div 6 = 4$.

Note that the example 48 ÷ 1.6 was deliberately chosen so that the number of digits after the decimal point was *not* the same in numerator and denominator. If the examples used in class have the same number of digits after the decimal point (for example 4.8 ÷ 0.6), pupils can easily get the impression that you remove the decimal points in both numbers and proceed. Then they make errors like 6.72 ÷ 0.8 = 672 ÷ 8.

Topic 24 Indices and standard form

1 Introduction

Indices provide an effective way of writing large numbers. The population of the world is about 6 000 000 000, a very long number. Using standard form this is 6×10^9. (Note that in large numbers there is a space after every three numbers to make them easier to read. Four figure numbers such as 2003 do not have a gap. Numbers with five digits or more have a space, for example, 23 456.)

Pupils need to learn the meaning of indices, the rules governing their use, and how to write numbers in standard form.

2 Squares and cubes

The first indices that children will meet are **squares** and **cubes** because these can be related to real-life situations.

$5^2 = 5 \times 5 = 25$ is read as '5 squared'
(for example,
5×5 is the area of a square of side 5 cm).

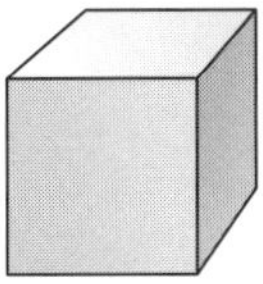

$6^3 = 6 \times 6 \times 6 = 216$ is read as '6 cubed'
(for example,
$6 \times 6 \times 6$ is the volume of a cube of side 6 cm).

It is quite useful if pupils know the squares and cubes of the first few whole numbers.

3 Meanings of indices

The pupils have learnt that 3×3 is written 3^2 (3 squared) and $3 \times 3 \times 3$ is written 3^3 (3 cubed). The teacher can extend this by using examples of larger **powers**.

For example, $3 \times 3 \times 3 \times 3 \times 3$, five 3s multiplied together, is written more briefly as 3^5. This is said as '3 to the power of 5' or '3 to the fifth'. 3 is called the **base** and 5 is called the **index**. The plural of index is **indices.**

The powers of 2 are important. It is useful if the pupils can recall some of them quickly.

$2^2 = 4 \quad 2^3 = 8 \quad 2^4 = 16 \quad 2^5 = 32 \quad 2^6 = 64$

4 The rules of indices

The rules are easily established from whole number examples.

For example,

$2^3 \times 2^4 = (2 \times 2 \times 2) \times (2 \times 2 \times 2 \times 2) = 2 \times 2 \times 2 \times 2 \times 2 \times 2 \times 2 = 2^7$

So $\quad 2^3 \times 2^4 = 2^{3+4} = 2^7$

A few examples like this will establish the rule $a^m \times a^n = a^{m+n}$.

Also, $2^7 \div 2^4 = \dfrac{2 \times 2 \times 2 \times 2 \times 2 \times 2 \times 2}{2 \times 2 \times 2 \times 2} = 2 \times 2 \times 2 = 2^3$. (by cancelling)

So $2^7 \div 2^4 = 2^{7-4} = 2^3$.

A few examples like this will establish the rule $\boldsymbol{a^m \div a^n = a^{m-n}}$.

Consider also, $(2^7)^3 = 2^7 \times 2^7 \times 2^7 = 2^{7+7+7} = 2^{7 \times 3} = 2^{21}$.

A few examples like this will establish the rule $\boldsymbol{(a^m)^n = a^{mn}}$.

The other rules follow from these.

For example:

Since $a^m \div a^n = a^{m-n}$, $a^m \div a^m = a^{m-m} = a^0$.
However, $a^m \div a^m = 1$ since any number divided by itself is 1.
So we have established the rule $\boldsymbol{a^0 = 1}$.

$2^4 \div 2^7 = 2^{-3}$ but $2^4 \div 2^7 = \dfrac{2^4}{2^7} = \dfrac{1}{2^3}$.
So, in general: $\boldsymbol{a^{-n} = \dfrac{1}{a^n}}$.

5 Standard form

Write a number such as 60 000 on the board. Ask pupils to write it first as a number multiplied by powers of ten, and then using indices.

$60\,000 = 6 \times 10\,000 = \mathbf{6} \times \mathbf{10^4}$
$64\,000 = 6.4 \times 10\,000 = \mathbf{6.4} \times \mathbf{10^4}$
$6000 = 6 \times 1000 = \mathbf{6} \times \mathbf{10^3}$
$6420 = 6.42 \times 1000 = \mathbf{6.42} \times \mathbf{10^3}$
$642 = 6.42 \times 100 = \mathbf{6.42} \times \mathbf{10^2}$
$642\,000 = 6.42 \times 100\,000 = \mathbf{6.42} \times \mathbf{10^5}$

This is the **standard form** (sometimes called scientific notation) of the numbers. They are written as a number between 1 and 10 multiplied by a power of 10, in the form $a \times 10^n$, where $1 \le a < 10$.

Standard form is much used in science for writing very large (and very small) numbers. The pupils have to learn how to change ordinary numbers into standard form. The teacher can use examples.

$60\,000 = 6 \times 10^4$
Changing 60 000 to 6 moves the digits 4 places.

$64\,000 = 6.4 \times 10^4$
Changing 64 000 to 6.4 moves the digits 4 places.

$6420 = 6.42 \times 10^3$
Changing 6420 to 6.42 moves the digits 3 places.

$642 = 6.42 \times 10^2$
Changing 642 to 6.42 moves the digits 2 places.

$642\,300 = 6.423 \times 10^5$.
Changing 642 300 to 6.423 moves the digits 5 places.

The power of 10 is the same as the number of the places the digits have moved. If the digits move to the right the power is positive. If the digits move to the left the power is negative.

6 Square roots and cube roots

Children should also learn about square and cube roots. ('Root' can be a puzzling word, as it has nothing to do with the roots of plants.)

$8^2 = 8 \times 8 = 64.$

Another way of stating this is to say that the **square root** of 64 is 8.

This is written as $\sqrt{64} = {}^{+}8$ and sometimes as $\sqrt[2]{64} = 8$.

Note that $\sqrt{64} = {}^{+}8$. It is true that $({}^{-}8) \times ({}^{-}8) = 64$ but the square root of a positive number is defined as always positive.

4 cubed, $4^3 = 4 \times 4 \times 4 = 64.$

This can also be stated as 'the **cube root** of 64 is 4', and written: $\sqrt[3]{64} = 4$.

Topic 25 Percentages

1 Introduction

Percentages are another way of writing numbers. % means 'per hundred'.

For example, 1% means $\frac{1}{100}$ or 0.01 ($\frac{}{100}$ has become %)

57% means $\frac{57}{100}$.

Pupils can be introduced to percentages by using a 100 square.

For example, this diagram shows 55% ($\frac{55}{100}$) of the square shaded.

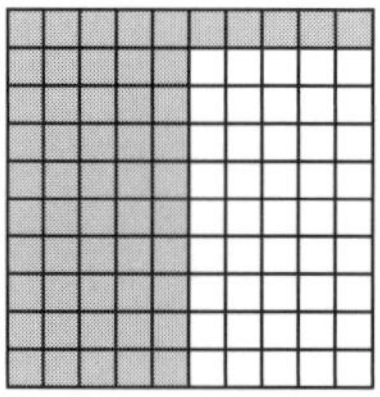

Changing a percentage to a fraction follows from the meaning of a percentage.

For example, 40% means $\frac{40}{100} = \frac{2}{5}$.

Changing a fraction to percentage requires us to multiply by 100% ($\frac{100}{100}$ or 1).

For example, $\frac{3}{8} = \frac{3}{8} \times 100\% = 37.5\%$.

2 How to solve percentage problems

There are a few basic types of problem that pupils will meet.

- **Find a percentage of a quantity.**

  For example find 35% of 80 cm. There are two ways to find the solution.

  (i) Write the percentage as a fraction and calculate the amount:

  $\frac{35}{100} \times 80 = 28$ cm

  (ii) Write the percentage as a decimal and calculate the amount:

  $0.35 \times 80 = 28$ cm

- **Express one quantity as a percentage of another.**

  For example,

  (i) What percentage is 60 cm of 4 m?

  First, the quantities need to be in the same units (change them both to cm).

  As a fraction it is $\frac{60}{400}$ which is $\frac{60}{400} \times 100\% = 15\%$

  (ii) What is 45 as a percentage of 60?

  45 is $\frac{45}{60}$ of 60, so $45 = \frac{45}{60} \times 100\% = 75\%$ of 60

  A common error: Because understanding the English is tricky, some pupils would calculate 45 as a percentage of 80 by writing $\frac{45}{100} \times 80$, which is 45% of 80.

- **Percentage increase or decrease.**
  For example,
  a population of 2000 increases by 8%. What is the new population?
  Again, there are two methods:
  (i) Calculate the increase and add it to the original amount:
  The increase is 8% of 2000 = $\frac{8}{100} \times 2000 = 160$
  The new population is 2000 + 160 = 2160
  (ii) An 8% increase means the population is now 108% of the original amount, or 1.08 times the original amount.
  $1.08 \times 2000 = 2160$
  This approach is much quicker, but more difficult, conceptually, for the pupils.

3 Discount, profit and loss

Discount

When items are bought in large quantities, the price is often reduced. Discounts are often expressed using percentages and are commonly set as a problem involving money.

For example: In a sale, all the goods in a shop are being sold at a 20% discount. What is the sale price of a dress whose original price was $28?

Discount = 20% of $28 = $5.60
Sale price of dress = $28 – $5.60 = $22.40
Or: New price = 80% of old price = 0.8 × $28 = $22.40

Profit and loss

These are often expressed as percentages. **Profit** is the **increase** of the selling price above the cost price and **loss** is the **decrease** of the selling price below the cost price. The percentage profit (or loss) is the profit (or loss), expressed as a **percentage of the cost** price.

For example: A shopkeeper buys radios for $60 each and sells them for $87 each. What is the percentage profit?

profit = difference between the selling price and the cost price
= $87 – $60 = $27

$$\text{percentage profit} = \frac{\text{profit}}{\text{cost price}} \times 100\% = \frac{27}{60} \times 100\% = 45\%$$

The shopkeeper has some radios left and sells them in a sale for $54. What is the percentage loss?

loss = cost price – selling price = $60 – $54 = $6

$$\text{percentage loss} = \frac{\text{loss}}{\text{cost price}} \times 100\% = \frac{6}{60} \times 100\% = 10\%$$

Activity

Find examples of the use of percentages by shopkeepers in their advertising.

Topic 26 Algebra

1 Introduction

Many pupils find algebra very difficult. They find letters, as opposed to numbers and words, very difficult to cope with. Algebra is difficult because it is not visual and it is complex.

One problem that pupils have is that they think of letters as shorthand for words. For example, if g is the number of girls in a class and b is the number of boys, then some pupils think that the expression $4g + 3b$ is short for 4 girls and 3 boys. This is wrong. $4g + 3b$ means $4 \times$ the number of girls $+ 3 \times$ the number of boys.

2 The idea of an unknown

Many books refer to a letter 'standing for' an unknown number. This is a difficult idea — what does 'stand for' mean — is it different to sitting? It is better to say, 'We will call the number n.' (It is common to type or print the letters in algebra in *italics*. The choice of letter used is unimportant.)

Develop algebra from simple examples.

- I have 3 sweets, my brother has 4 sweets. Together we have $3 + 4 = 7$ sweets.
- I have 5 sweets, my brother has 4 sweets. Together we have $5 + 4 = 9$ sweets.
- I have n sweets, my brother has 4 sweets. Together we have $n + 4$ sweets.

Here n can be any number of sweets. If I have 9 sweets, replace n by 9 in '$n + 4$' and it becomes $9 + 4 = 13$. Link the algebra with language: '$n + 4$' is 4 more than n.

Early situations that give rise to the idea of an unknown and simple formulae should involve only one operation.

Multiplication	1 person has 2 ears, so n people have $2 \times n = 2n$ ears. (Explain to pupils that $2n$ is short for $2 \times n$.) 1 football team has 11 players, so x football teams have $11x$ players.
Division	If 4 people can sit at one table, n people will need $n \div 4$ tables.
Subtraction	The number that is 3 less than n is $n - 3$. The difference between x and 7 is $x - 7$.
Addition	My sister is 3 years older than me. If I am x years old, she is $x + 3$ years old.

3 Formulae

Pupils meet quite a few formulae in mathematics. For example, $A = lw$, $A = s^2$, $P = 2(l + w)$. It is usually better to keep the formulae

> ***Activity***
>
> Think of some more situations that give rise to simple formulae.

in words for as long as possible:
area = length × width, area = length of side × length of side,
perimeter = length + width + length + width.

Substituting into a formula

For example, find the value of $3x + 5$ when $x = 4$. The letter is replaced by a number.
Start with simple expressions and build up to more complicated ones.

- Substitute into one-operation formulae. Evaluate $x + 5$ when $x = 2$.
- Substitute into two-operation formulae. Find $4x + 5$ when $x = 2$, and $20 - 3y$ when $y = 4$.
- Substitute into formulae with two unknowns. Evaluate $4x + 5y$ when $x = 2$, $y = 3$.

A game: 800 m (for 2 to 5 players)

Equipment:
A picture of a running track like the one below.
9 cards: three cards with 1, three cards with 2, three cards with 3 on them.

The track is in sections, each of which has a formula in it. Shuffle the cards and place them face down in the space in the middle of the track.

Each player has a counter or small object and puts it on the space with '$x + 1$ START' written on it. The first player picks the top card from the pile of cards and substitutes the number on it into the formula of the section on which his/her counter is placed. The numerical value is the number of spaces he/she moves his/her counter forwards. For example, if a player picks up a 2 when on the $x + 1$ space, he moves $2 + 1 = 3$ spaces clockwise. The players take it in turns to pick up the next card from the top of the pile and

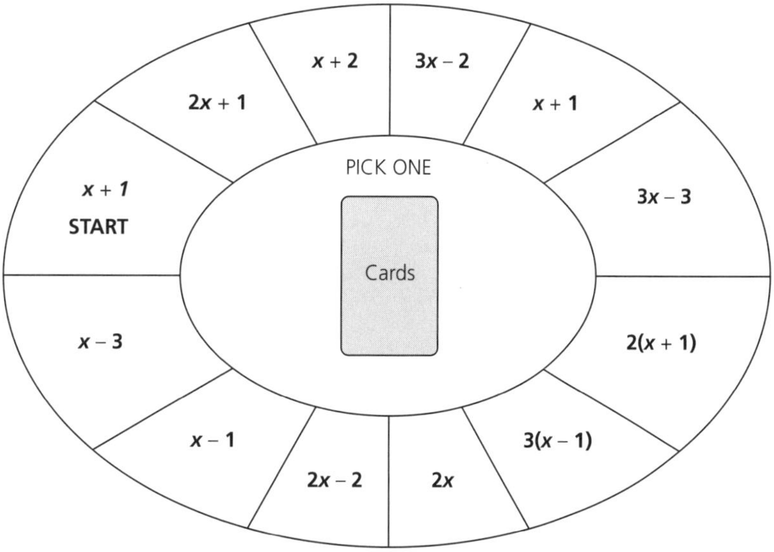

substitute the number into the formula that their counter is on. The result tells the player how many places to move. The winner is the first one whose counter goes round twice (hence the name: 800 m). This game gives good practice in substituting in simple formulae.

4 Simplifying expressions

For example, simplify $3a + 5b - 2a + 4b$.

Letters are operated upon without it being necessary to know what they stand for. Pupils have to learn what **terms** are, what an **expression** is and what **like terms** are. A 'term' is any product of letters and numbers, such as $4xy^2$ or $3p \div q$. An 'expression' is any sum or difference of terms, such $4x^2 - 3xy$, 'like terms' are terms containing the same letters, such as $7b$ and $4 \times 3b$, and $2x^2$ and $45x^2$.

Beware: Teachers often say that expressions such as $3a + 4b$ cannot be **simplified** because the terms are different. They use the example 4 apples + 3 apples is 7 apples, but 3 apples and 4 bananas cannot be simplified. Apples and bananas cannot be added. However, they then go on to say that $3a \times 4b = 12ab$. This is very odd – apples and bananas cannot be added but they can be multiplied! Adding apples and bananas is like putting them in the same fruit bowl. Multiplying them makes no sense at all.

The correct reason that $3a + 4b$ cannot be simplified is that **we do not know what numerical values *a* and *b* have. When we know that, we can go further**.

In some questions, letters are ignored. Once the letters have been noted, nothing further has to be done with them when answering the question. For example in 'Add 4 to $n + 7$', 4 is added to 7 and the answer is $n + 11$. The n has not been changed.

Topic 27 Solving equations

1 Introduction

Pupils should first solve equations involving only one operation, such as $4 + n = 20$, or $4n = 20$, or $n - 4 = 20$, or $\frac{n}{4} = 20$. At this stage, all equations should have whole number answers. Later they can solve equations that combine two procedures such as $2n + 4 = 20$, and $7x + 13 = 27$. The letter in each equation represents a specific unknown number, which is operated on and evaluated.

There are several methods of solving equations.

2 Solving equations by substitution

The pupil tries various values until he finds the correct one. This method works well if the answer is a small whole number but not if it is a fraction.

For example, $7x + 13 = 27$

Try $x = 3$. Then $7x + 13 = 21 + 13 = 34$, too big.
Next try $x = 2$, then $7x + 13 = 14 + 13 = 27$, correct,
so $x = 2$ is the solution.

3 Solving equations by 'doing the same to both sides' (DSBS)

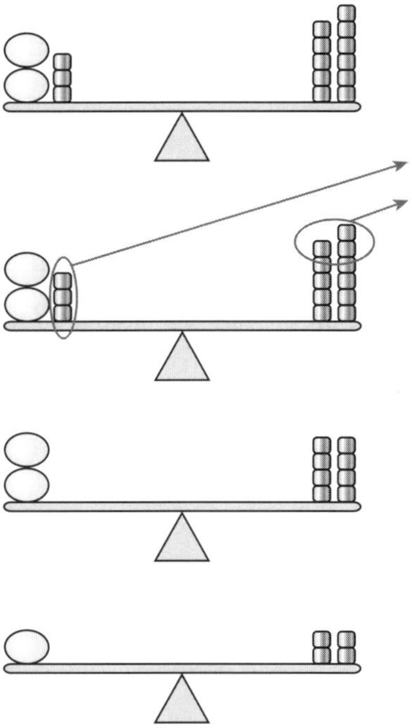

The equation $2x + 3 = 11$ illustrates the process.
Two bags have the same number of identical weights in them. With 3 weights, they balance 11 weights.
By removing weights, it is possible to find out how many weights are in the bags without opening them.

$$2x + 3 = 11$$

Three is subtracted from each side. $\quad 2x + 3 - 3 = 11 - 3$

$$\Leftrightarrow \quad 2x = 8$$

Each side is halved. $\quad \frac{1}{2} \times 2x = \frac{1}{2} \times 8$

$$\Leftrightarrow \quad x = 4$$

The balance is used to illustrate the process. Both sides must be changed in the same way if the balance between the two sides is to be maintained. The equals sign here does not mean the usual 'this is the result of', but a more sophisticated, general 'this is the same as'.

Activity

Make a balance to illustrate this.

You will need three pieces of wood joined as in the diagram. The base and the upright have to be rigid. The top piece has to be able to swing. At the ends of the top piece are hooks to hold several strings holding two flat tin lids. The weights are placed on the lids.

The balance illustrates effectively the idea that what is done to one side of an equation (add, subtract, multiply or divide) has to be done to the other side if the two sides are to remain equal to each other. This method is quite good preparation for later, more complicated, work but it may seem unnecessarily complicated to a young child.
For example, if $n + 4 = 7$ then n must be $7 - 4 = 3$.
Writing $n + 4 - 4 = 7 - 4$ is very long winded.

Give pupils practical experience using this method.
Illustrate $n + 4 = 7$ with 7 weights balancing 4 weights plus 3 hidden in a bag. Taking 4 weights from both sides leaves the weights in the bag balancing 3 weights so the unknown number of weights is worked out without looking in the bag.
(Note the use of $\Leftrightarrow$. This sign means that each equation implies and is implied by the other connected to it with the sign.
$2x + 3 = 11 \Leftrightarrow 2x = 8$ means that $2x + 3 = 11$ implies $2x = 8$ and $2x = 8$ implies $2x + 3 = 11$.)

4 Solving equations by 'change side, change operation' (CSCO)

Addition on one side of an equation has the same effect as subtraction on the other. (Think about the scales again.)
Multiplication on one side is the same as division on the other.

$4 + 5 = 9$ is 'balanced', so is $4 = 9 - 5$
If $n + 4 = 7$, then $n = 7 - 4$ is also true.

$4 \times 5 = 20$ is 'balanced', so is $4 = 20 \div 5$
If $4n = 20$, then $n = 20 \div 4$ is also true.

Look at the previous example again:

$2x + 3 = 11 \Leftrightarrow 2x = 11 - 3$ [+3 on the left has changed to – 3 on the right]

$\Leftrightarrow 2x = 8$

$\Leftrightarrow x = 8 \div 2$ [× 2 on the left has changed to ÷ 2 on the right]

$\Leftrightarrow x = 4$

The CSCO strategy is based on the DSBS method. It is popular because it is short. However it can cause problems if it is not properly understood.

- Pupils must change over numbers **and** addition and subtraction signs. For example they may write:

  $30 = x + 7$, so $30 + 7 = x$
  or $4x - 5 = 19$, so $4x = 19 - 5$

- They may separate numbers and variables. For example they may write:

  $8k = k + 14$, so $8 - 14 = k - k$

- They may reverse operations. (Pupils do this to make a simpler answer.) For example they may write:

  $x + 20 = 14$, so $x = 20 - 14$

  $4x = 2$, so $x = 2$
- They may treat subtraction as commutative. For example they may write:

  $20 = 7 - x$, so $20 - 7 = x$

For these reasons, it is important that pupils learn CSCO from the DSBS method. (Some pupils learn to 'Change side, change sign'.) This is confusing.

$2x = 4 \Leftrightarrow x = 4 \div 2$

The 2 has changed sides but its sign (positive / negative) has not changed.

5 Solving equations by using inverse operations

You can teach equations by using the idea of inverse operations. For example, multiplying by 2 is undone by its inverse, dividing by 2; adding 2 is undone by its inverse operation, subtracting 2, and so on.

Using the idea of a balance,

$n + 4 = 7$ becomes $n + 4 - 4 = 7 - 4$ (DSBS).

Adding 4 is undone by subtracting 4.

So $n + 4 = 7 \Leftrightarrow n = 7 - 4$ (CSCS).

Illustrate the idea of inverses by considering what happens when the operations are reversed. For example, suppose you wrap a box in purple, then blue, then yellow, then green paper. To take the paper off we remove green, then yellow, then blue, then purple paper. In the morning, I put on socks and then shoes. In the evening, I remove shoes and then socks. An equation such as
$3x + 7 = 19$ can be undone as well:

x → (× 3) → $3x$ → (Add 7) → $3x + 7 = 19$

Reversing the process:

4 ← (÷ 3) ← 12 ← (Minus 7) ← 19

The solution is $x = 4$.

6 The equals sign

The '=' sign can cause problems for pupils because it is used in several quite different ways. In algebra it is used:

- in **equations**, such as $3x + 1 = x + 7$ — These have one solution.
- in **identities**, such as $2(x + 5) = 2x + 10$ — This is true for all values of x.

- in **functions** such as $y = 3x + 4$ True for an infinite set of pairs.
- in **formulae** such as $V = l\,w\,h$. True for an infinite number of sets.

They may also meet the symbol in other contexts where it is used as a shorthand:

- Scale: 1 cm = 1 m here '=' means 'represents'
- Ans. = 25 m, My height = 1 m 73 cm here '=' means 'is'
- 4 men = 52 hours here '=' means 'take'

Children first meet the sign in sentences such as '4 + 2 = 6'. Here '=' means 'makes'. The equals sign indicates the result of an operation to solve a problem. In Primary school '=' is not usually thought of as indicating equality. It is introduced as a shorthand, '4 + 2' is a 'problem' or a 'question' and 6 is the 'answer'.

Young children will find a sentence such as '3 = 8 – 5' or '3 = 3' puzzling. The first sentence is 'the wrong way round'. The second contains no operation.

In algebra, the child has to develop a much more sophisticated understanding of the symbol. An equation such as $5x + 7 = 22$ fits the pupil's earlier perception of the meaning of the symbol. When the equation is reversed, as in $27 = 8x - 13$, some children can have difficulties. Similarly, an equation such as $8 = 6x$ is often more difficult for children to solve than $6x = 8$.

Topic 28 Sets

1 Introduction

The topic of set theory is often included in a syllabus because it unifies various aspects of maths. It has in recent years become less significant as a Primary school topic but it does remain in many syllabuses. Teaching this topic offers many opportunities for interesting activities.

2 Describing a set

A **set** is just a collection of things. Mathematically speaking we say a set is a **well-defined** group of objects. That means we can tell whether anything is or is not a member of the set. For example, the set {some pets} is not a well defined set because we do not know precisely which pets are in the set. The set {the pets of pupils in Grade 5 in our school} is well defined.

Usually the **elements** in a set have something in common — such as all being cakes, or birds, or vegetables, etc. A useful exercise is to sort a collection of items into sets. For example, sort the children into the set of those who have brothers or sisters in the school, the set of those who are 10 years old, the set of boys, and so on. Collect objects and then the pupils can sort them by their shape, or by colour, or by size. Pictures of vehicles can be sorted by category (car, van, and lorry) or by colour.

These pictures can be sorted into different sets by size, as round or not round, shaded or not shaded, big or small.

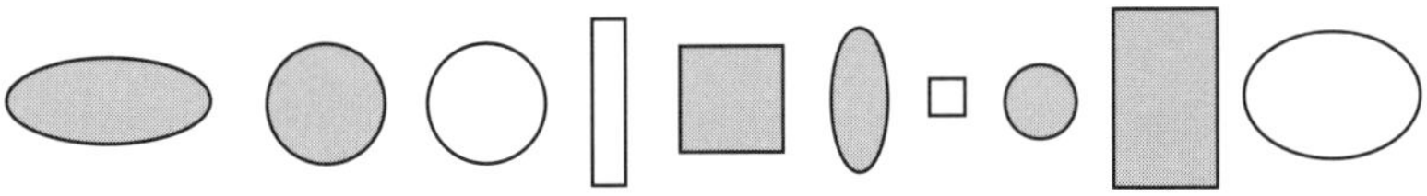

These can be sorted by shape (triangle, square, circle), size (small, medium, large) or shading. (These are like Dienes' **attribute blocks**.)

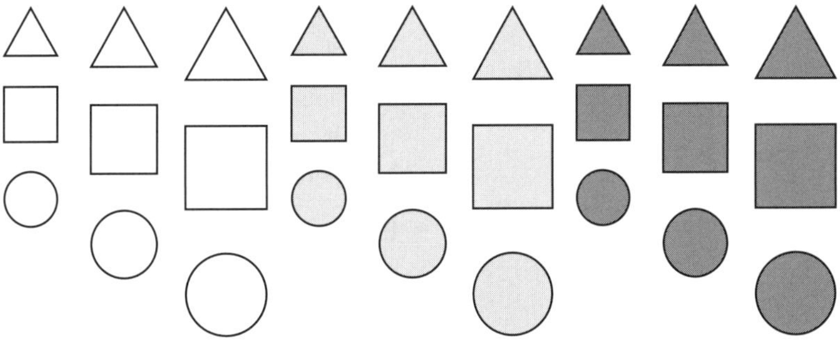

Activity

Use card to make copies of these shapes. The pupils can use them to sort into different sets. If possible, make them with card of two different thicknesses.

A game: Guess the set (for the whole class)

Arrange the class in rows and columns. Each column is a team. One pupil goes to the front of the class and is given a card by the teacher with the description of a set on it, for example 'fruits' or 'triangles'. The pupil has to draw something that would be in the set. The rest of the class have to guess the name of the set. The pupil can keep drawing objects that would be in the set until somebody guesses correctly. The person who guesses correctly scores one point for his/her team. Teams take it in turns to have someone to do the drawing.

A game: Five in a set (for 2 to 4 players)

This game needs 80 cards with pictures on them. Each player starts with five cards and has to try to make a set of pictures that are connected in some way (all animals, all brown, all have wheels, etc.). The starter picks up a card, decides which cards to keep and discards one to the bottom of the pack. The players play in turn. When a player has five cards that make a set, he must show them to the other players and name the set.

3 The language of sets

Elements or members of a set

Pupils need to use the language of sets when learning about them. Ask pupils to list the **elements** or **members** (use both terms frequently) of various sets — the days of the week beginning with T, the class members with spectacles, the odd numbers from 0 to 10, the vowels and so on. Pupils need to use the { } notation to write sets, as in {2, 4, 6}. Pupils should be able to describe sets by:

- listing all the elements, {1, 2, 3, 4, 5, 6}, (Ghana, Nigeria, Sierra Leone}
- listing some elements and using dots, {1, 2, 3,.......,19, 20}
- listing some elements for infinite sets {1, 2, 3,....}, {2, 4, 6, ...}
- describing the elements, {the whole numbers from 1 to 6}, {the pupils of Uganda}, {the countries of Africa}.

Introduce set theory by using real objects such as fruit, stationery, boxes and containers. Ask children how to group things. Choose things that they are familiar with, such as dress colour, shoe size, birthday month. Collect pictures of animals to sort by the number of legs (0, 2, 4, 6 or 8; discuss why they are all even numbers.)

Equal sets

Two sets are **equal** if they have the **same** members. The order in which the members are written does not matter.
For example, {1, 3, 5, 7} = {7, 5, 3, 1}.

Equivalent sets

Pupils sometimes think that sets are equal if they have the same number of elements. For example, A = {1, 3, 5, 7} and B = {2, 4, 6, 8} have the same number of elements, 4, but they are not equal.

$n(A) = n(B)$ but $A \neq B$.

Sets with **equal numbers of elements** are called **equivalent**.

Disjoint sets

Two sets that have **no elements in common** are called **disjoint** sets.

4 Venn diagrams, intersection and union

Use examples using class members to introduce all these ideas. Ask a set of girls to come to the front. Ask a set of boys to stand. They will stand quite separately. The two sets are disjoint. We can show this in a picture called a **Venn diagram**.

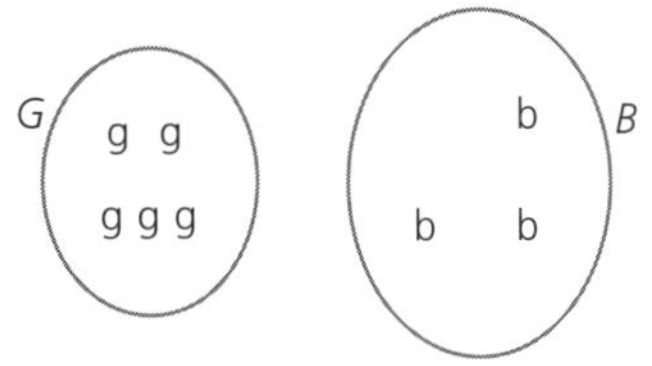

Venn diagram of the children

You can use string and put it in a loop on the ground. Ask the boys and girls to stand inside the loops. Use other class situations for further examples.

Sets can have common elements. For example, J = {boys whose names begin with J}, T = {tall children}. Ask the two sets to come to the front of the class. Ask set J to stand on one side of the class and set T to stand on the other. Where should a tall boy called Joseph stand? The loops will have to overlap, so that Joseph can stand in the middle and be in both sets. Then the tall boys whose names begin with J are in both sets. You can show this on a Venn diagram. (Use the pupils' names when you draw the diagram.) The overlap is called the **intersection** of the sets. In this case it is the set of tall boys whose names begin with J.

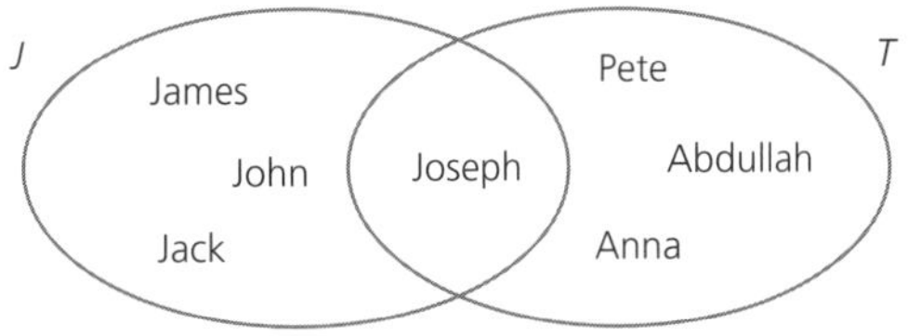

Use numerical examples to introduce the notation.

$A = \{1, 2, 3, 4, 5, 6\}$, $B = \{2, 4, 6, 8, 10\}$ and $A \cap B = \{2, 4, 6\}$

If we look at the set of **all** the children at the front of the class, we have the **union** of the two sets. It consists of all elements that are in one set, or the other, or both.

Using the numerical example above we would write:

$A \cup B = \{1, 2, 3, 4, 5, 6, 8, 10\}$.

Pupils need a lot of practice drawing Venn diagrams. Use examples from their lives, or from objects they are familiar with, or from sets of numbers, or from sets of letters.

5 The empty set

The empty set is the set that contains no elements. There is only one empty set, so do not talk about *an* empty set, only about **the** empty set, written ϕ or { }.

Introduce the empty set by asking pupils for:

- the set of people over 3 m tall
- the set of people over 200 years old
- the set of pupils in your school more than 30 years old
- the intersection of two disjoint sets, such as the boys and girls in the class.

All these sets have no elements. They are empty. Ask the pupils to suggest other examples. These can be very amusing!

6 Subsets

Some sets are completely 'inside' another. For example, choose the set of girls from the class. Select the tall girls. Ask them to stand in the middle of the set. The set of tall girls is a **subset** of the set of girls.

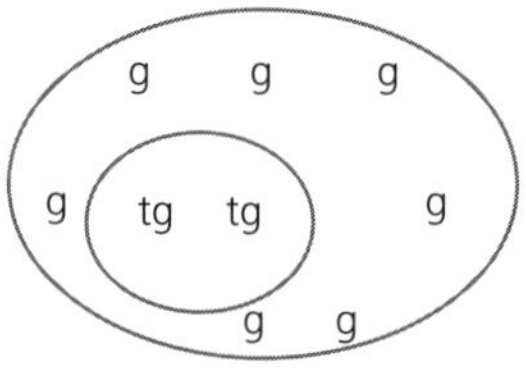

Give some numerical examples and introduce the notation.

$A = \{1, 2, 3, 4, 5, 6\}$, $B = \{2, 3, 5\}$
B is a subset of A, $B \subset A$.

In the Venn diagram, B is inside A.

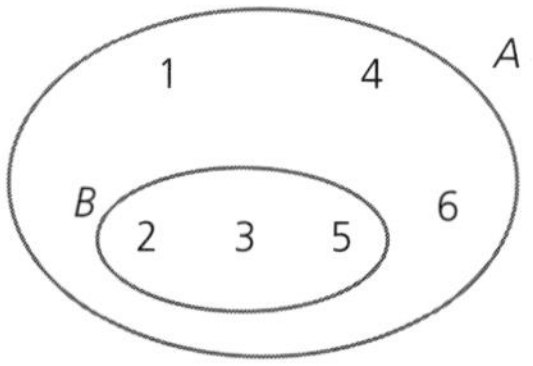

Make sure that the pupils understand that the empty set is a subset of every set.

Number of subsets

Every set is a subset of itself, and the empty set { } is a subset of every set. You can establish a formula for the number of subsets of a set by making a table like the one on page 149 of all the subsets of sets with different numbers of elements. (Leave the top two lines until last as they are tricky.)

Set	Subsets	Number of elements	Number of subsets
ϕ	ϕ	0	1
{a}	ϕ, {a}	1	2
{a, b}	ϕ, {a}, {b}, {a, b}	2	4
{a, b, c}	ϕ, {a}, {b}, {c}, {a, b}, {a, c}, {b, c}, {a, b, c}	3	8
{a, b, c, d}	ϕ, {a}, {b}, {c}, {d}, {a, b}, {a, c}, {a, d}, {b, c}, {b, d}, {c, d}, {a, b, c}, {a, b, d}, {a, c, d}, {b, c, d}, {a, b, c, d}	4	16

Adding another element doubles the number of subsets each time.

Following the pattern, it is easy to see that the set {a, b, c, d, e} with 5 elements has 32 subsets. The pupils should be able to say that the number of subsets is given by the powers of 2, and that a set with n members has 2^n subsets.

7 Universal sets

When we say that a set is the **universal** set, we are merely stating what is the collection of things that we are referring to at that particular time. Choosing a universal set is a good class exercise. When we ask questions about what the universal set is given a couple of sets, there could be lots of different possible answers. For example, suppose A = {1, 3, 5, 7, 9}, B = {2, 4, 6, 8}, the universal set could be 'natural numbers less than 10', but it could also be 'whole numbers less than 10', or 'integers'. Because there are so many possible answers it is better not to put questions in exams that ask pupils to name a universal set. Pupils often answer questions like these by merely repeating the sets given in the question. For example, to answer the request
'Name a universal set for A = {1, 3, 5, 7, 9}, B = (2, 4, 6, 8}',
many pupils will write {1, 3, 5, 7, 9, 2, 4, 6, 8}. This is a correct answer but it is also the union of the sets and it is not clear that they have understood what 'universal set' means.

Usually the universal set is stated and the children have to decide which members belong to the universal set. The children can be shown a universal set of various shapes and asked to sort it in various ways – by colour, by number of sides, and so on. They are, in effect, finding subsets.

Activity

Collect objects suitable for use in classification exercises. These could include:

- **Ribbons** Take ribbons out of a box and group by colour. Count the number of each colour.
- **Jars** Group them according to shape, colour, contents, and type of lid.
- **Sports** Use pictures and sort into sets — ball games, team games, etc.

Prepare boxes full of objects. Pupils can work in groups to discuss/classify the objects into sets.

Section C Mathematics topics — handling data

Topic 29 Collecting and organising data

1 Introduction

Pupils need to learn to understand statistics. They need to be able to interpret, and not be misled by, statistics presented in newspapers, advertisements, and so on. Statistics also give pupils an opportunity to apply arithmetic in a meaningful way. There are several stages to statistics work in Primary school. These are:

- collection and organisation of data
- displaying the data in tables and diagrams
- analysing and discussing data.

This topic looks at the collection and organisation of data.

2 Collecting data

Too often statistics is taught using information that is in the textbook and is not very interesting for the pupils. It is much more interesting for them if the information they use in class is:

- information about them
- information that they have collected themselves.

Collecting the information should be enjoyable. The information should be of some interest to the pupils, and not too difficult to collect. The process of collection should also be educational. As they

record the information to bring to the classroom, the pupils are measuring, counting and writing in an organised way.

3 An introduction using birthdays

Preparation by the teacher Prepare pieces of paper, one for each pupil, and organise the collection of small boxes or cubes. Put labels for the 12 months January, February, March, etc. in order, in a line at the front of the class. Write the names of the months across the bottom of the board.

Preparation by the pupils Each child draws a picture of himself/herself on the piece of paper given out by the teacher. Each child also needs a small identical cube or box. Empty matchboxes will do.

The activity The pupils do three things:

- They place their picture in the column for their birth month on the board. The diagram is a **pictograph** showing the birth months of the whole class.
- Each child then places his/her box in the pile for the correct month on the teacher's desk. This creates a **bar chart** of the same information.
- They sit or stand in front of the month of their birth. This makes a pictograph — a living one.

4 Information from within the class

The children have a lot of information from their everyday lives that that they can use in the classroom:

- numbers of brothers and sisters, aunts and uncles, absentees per day
- favourite colour, football team, fruits, etc. of class members
- heights of pupils, lengths of arms, reaction times
- the number of glasses of water each pupil drinks in a day.

You can also use activities:

- Look at bags of objects of mixed colours (count the number of different coloured things)
- Show pupils a picture with ten different animals in it for twenty seconds. Then ask them to write the names of the animals they saw on a piece of paper. Collect the papers and see how many animals the pupils could remember.
- Standing jump: pupils make a standing jump. Pupils measure and record the distances. Measuring tasks like this lead on to working with **grouped data**.

Try to involve as many pupils as possible in all of these investigations.

5 Information from outside the class

The children can collect information from many sources outside the classroom. Examples include:

- data about the school, the number of pupils in each year, the number of boys and girls in each year
- counting the number, colours and types of cars going past the school
- collecting information about the trade, industry and population in your village, town or country
- collecting information about the environment, pollution, and so on
- collecting geographical information
- looking at health issues, for example, the data on cigarettes and lung cancer.

Caution: Data collecting activities outside the classroom need careful supervision. The teacher has to be certain that what his/her pupils are doing is safe.

6 Activity regions

One way to organise a lesson of data collection is to arrange for different groups in different parts of the classroom to be doing different things. For example, a class is divided into 6 groups. Each group has a different task to do that has been written on a sheet of paper. Pupils in a group will then carry out the activity. For some of the activities they will have to question the other pupils. After completion, the teacher will choose a group member to present a report to the class. This report could involve producing a pictogram, a bar chart or a pie chart.

Activities for the groups could include the following.

- Find out the birth months of pupils in the class.
- Find the number of siblings of everyone in the class.
- If they roll a die 120 times, record the number of times that 1, 2, 3, 4, 5 and 6 come up.
- Find out which is the most commonly used letter of the alphabet by counting the letters a, b, c, ... in a passage from a book.
- Find out the favourite colour, football team, etc. of pupils in the class.
- Place some coloured beads in a glass jar. Ask pupils to estimate how many there are in the jar. Record their guesses and then count them.

Before each group presents its report, discuss with the class what results you expect them to come up with.

7 Tallying

Tallies help to make the collection of data easier. For example, suppose some pupils were writing down the types of vehicles going past the school. They could write down: *Lorry, bus, car, bus, car, car, lorry, car, bicycle, car*, etc. but this would be tedious and they would be likely to make mistakes.

A better idea is to use a tally chart. Pupils think about the vehicles they are likely to see and prepare a blank chart. (If they see any additional vehicles these can be added to the chart.)

Each time a vehicle passes they put a mark / in the appropriate row. When a fifth vehicle of a particular type passes they record it by striking through the previous 4 tallies ~~////~~. This makes it easier to count the totals.

Finally, the totals are recorded.

At the start:

	Tallies	**Total**
Lorry		
Bus		
Car		
Bicycle		

Part way through:

	Tallies	**Total**
Lorry	//	
Bus	//	
Car	~~////~~ //	
Bicycle	/	

At the end:

	Tallies	**Total**
Lorry	////	4
Bus	///	3
Car	~~////~~ ////	9
Bicycle	////	4

Topic 30 Displaying data on pictographs, bar charts and line graphs

1 Introduction

When data has been collected, it must be displayed in a way that people can understand it. A list of the birthdays of the pupils in a class is not very informative. Such a list may look like this:

October 11th	September 26th	July 28th	November 5th
November 23rd	December 14th	May 2nd	December 2nd
January 3rd	February 21st	September 27th	May 9th
April 12th	May 7th	July 7th	August 3rd
September 26th	October 15th	May 28th	May 27th
September 3rd	December 27th		

This list needs to be organised in a table so that it can be understood:

Month	Jan	Feb	Mar	Apr	May	Jun	Jul	Aug	Sept	Oct	Nov	Dec
No. of birthdays	1	1	0	1	5	0	2	1	4	2	2	3

This information can then be displayed in a pictograph or bar chart.

2 Pictographs

The simplest charts are pictographs in which a picture of a person (or object) is used to represent a person (or object). Pupils can have fun making pictures to be used in pictographs.

Often the picture or symbol will represent more than one person (or object). Pupils should meet examples of this. If a picture represents two people, then one person is shown by half a picture. Pupils should see examples of pictographs where the symbol represents two, ten or more people/objects. They will need to work out how many half a symbol/quarter of a symbol represents. When the frequencies are large, and a symbol has to stand for many things, pictographs are not very accurate. However they are good for showing small frequencies, and for showing comparative sizes simply.

Show the pupils an example of a pictograph drawn on a large piece of card. Explain what the pictograph is about (it should have a title) and make sure all the pupils can see it very clearly. From the pictograph, you can ask a lot of questions by using the words '*how many?*', '*what?*', '*which?*', etc.

3 Bar charts

When the pupils know how to read a pictograph, the same information can be shown on a bar chart. Bar charts can be drawn vertically or horizontally. Demonstrate to the pupils how to draw a bar chart on the board based on the information given on the pictograph. A bar chart is easier and quicker to draw than a pictograph.

Use another example with the pupils, for example, *what is their favourite cartoon?* Put a list on the board and collect the results by marking tallies next to each item on the list. (It is best to provide a list of six or seven items for your pupils to choose from, or else you may end up with a chart with twenty items and only one or two tallies next to each of them.) Then draw the bar chart to show the information. Make sure the chart has a title, axes with labels and a scale if necessary, and is drawn neatly.

For example:

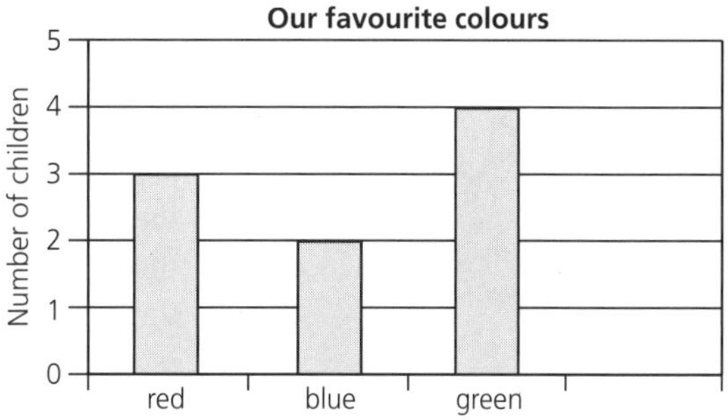

An activity that can be done in the classroom

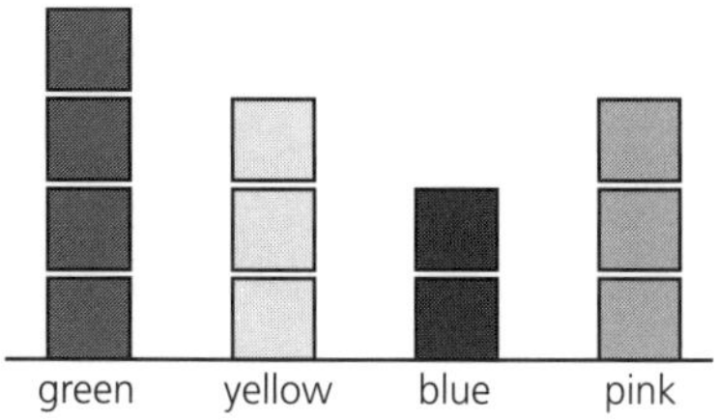

Give each pupil a piece of card with one of four different colours e.g. green, yellow, blue and pink (make sure you have different numbers of each colour). Then, ask each pupil in turn to come to the front and stick his/her card on the board. The cards will produce a bar chart. The pupils can then draw a bar chart from the information given on the board, giving their work a title and the axes the appropriate labels.

When pupils have produced pictographs and bar charts, they can be displayed in the classroom.

4 Line graphs

So far we have only looked at **discrete data** (data that involves counting), but much of statistics is about measuring. Data of this type is called **continuous data.** Continuous data can be shown on a line graph. Line graphs are an effective way of showing data recorded on a regular basis. For example, suppose the height in centimetres of a plant is measured each week as follows:

Week 1	Week 2	Week 3	Week 4	Week 5
1	5	12	15	17

The line graph would look like this:

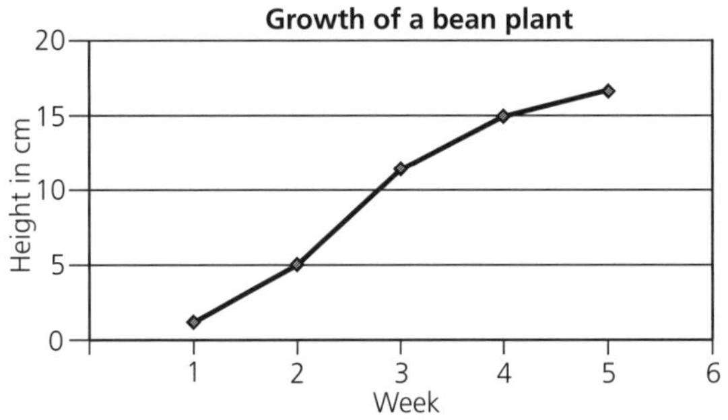

This shows graphically how the plant grew quickly during the first few weeks but slowed down in the later weeks. This could be because the plant was full size, or because of a water shortage.

Displaying data like this can be used to give pupils interesting scientific and mathematical experience. Other examples can be used:

- the reduction in the time for the world record for the 100 m, 200 m and other races over the years
- the increase in the world records for the high jump, long jump, javelin and other events
- the temperature at different times of the year, different times of the day, or in different countries
- the rainfall, hours of daylight, and so on, at different times of the year.

Activity

Collect data so that pupils can prepare pictographs, bar charts and line graphs.

Find out athletics data, rainfall data, hours of daylight data, and so on.

Topic 31 Mean, median and mode

1 Introduction

When data has been collected, it has to be interpreted. One way it can be interpreted is by trying to find a single value that can be thought of as being representative of the data as a whole. The representative value that is most commonly used is the **average**. Pupils learn about three averages. These are the **mean**, the **median** and the **mode**. They are sometimes called **central measures**. Pupils need to learn how to find them and about their relative advantages and disadvantages.

2 Mean

The mean of a set of scores is found by adding them together and dividing by the number of scores.

For example, the mean of the five scores 4, 6, 7, 4 and 9 is

$$\frac{4 + 6 + 7 + 4 + 9}{5} = \frac{30}{6} = 5.$$

This is the most common 'average' used in everyday life. When people use the word 'average' in conversation, they are usually talking about the mean. Meteorologists talk about the average temperature, cricketers talk about batting averages, and so on.

(In more advanced mathematics, this is called the arithmetic mean, to distinguish it from the geometric mean, $\sqrt[5]{4 \times 6 \times 7 \times 4 \times 9.} = 5.7$)

Finding the mean of some results

Many interesting mean values can be found. For example:

- The mean income of every person (income per head) in a country is found by dividing the total income of a country by the number of people in it. Find out this information for your country and other countries and get the pupils to calculate and comment on the results.
- The mean number of brothers and sister of the pupils in your class as compared with the mean number of brothers and sisters of their parents.
- The mean number of goals scored in each game by the best and other football teams, and the mean number of goals scored against them.
- The mean temperature and rainfall at different places in the country, in Africa, and in the world, and at different times of the year.
- The mean height of the pupils in your class.
- The mean of test or exam scores. This is data with which pupils are familiar. The analysis of the data can lead to interesting

questions. For example, suppose a pupil got the following marks in English and maths.

Subject	Pupil score	Class mean
Maths	48	42
English	66	68

In which subject did the pupil do better? The answer to this depends on what you want to compare. He scored 'better than average' in maths and 'worse than average' in English, this would suggest maths is his better subject. He scored more marks in English, so that would suggest English was his better subject. However, we also need to know if both tests had the same total marks, or if the marks we are given are a percentage — 66 out of 500 and 48 out of 50 leads to a different interpretation of the results!

Encourage your pupils to ask questions about the data they are given.

An activity for pupils

Divide the pupils into groups. Give each group a long piece of string. Ask them to show the mean height of the group using the piece of string but without a ruler. Do not tell them what to do. Get them to work out a solution for themselves. There are several different ways that they can use the string. This activity will help develop a good idea of the mean of some measurements and enable the pupils to practise problem-solving skills.

3 Median

The **median** is the *middle* score in a set of scores. This is easy to find if the number of scores is odd. The pupil has to put them in order and note the middle one. For example, scores 4, 6, 7, 4 and 9 in order are 4, 4, 6, 7, 9. The median is 6. When introducing this idea, make sure that you use an odd number of scores, because the median of an even number of scores is slightly more complicated. For example, what is the median of 4, 5, 6, 17, 18 and 19? The middle two numbers are 6 and 17. We say the median is the mean of these two numbers which is

$$\frac{6 + 17}{2} = \frac{23}{2} = 11.5.$$

A formula for the median is not needed at this stage.

4 Mode

The **mode** is the *most common* score in a set of scores. For example, the most common score in the scores 4, 6, 7, 4 and 9 is 4 which occurs twice. This is the mode. The mode is usually easy to find. Sometimes there is more than one mode. In these scores 4, 6, 7, 4, 9, 4, 6, 7, 6 and 9 there are three 4s and three 6s. 4 and 6 are the two modes.

Some pupils confuse the mode and its frequency. For example, suppose the frequency table of some scores is:

Score	2	3	4	5
Frequency	3	7	6	2

The modal score is 3, because that has occurred the largest number of times. Some pupils think that the mode is 7, the number of times that the most common score has occurred. It can help pupils to look at examples in which the information collected is not about scores or numbers, as in this table of pupils' favourite colours.

Favourite colour	**Red**	**Blue**	**Green**	**White**
Frequency	3	7	6	2

The most popular choice clearly has to be a colour and not a frequency.

5 Comparing the three 'm's

The mean is best suited to be used for data that is *evenly spread.* The advantage of using the mean as average is that it takes into account *all* the values in the set of data. It is easy to calculate. The total for a given number of values can be calculated from it.

A disadvantage is that finding it involves a calculation. Moreover, extreme values can distort it. For example, if the scores in a test were 14, 15, 16 and 99, the mean score is $\frac{144}{4} = 36$ which is very much larger than all but one of the scores.

The median is easy to find if the data is ordered, and it is *not so affected by extreme values*. It is a useful measure when there are extreme scores. The median of the numbers 14, 15, 16 and 99 is 15.5, which is close to most of the scores. The disadvantage is that it does not use all the scores. The median of 14, 15, 16 and 99 is the same as the median of 3, 15, 16 and 17.

The mode can be used for non-numerical data (such as which shirt colour is the most popular). It is very easy to find. It is *not affected by extreme scores*. Its disadvantage is that it does not use all the values and there may be more than one mode. For example, 14, 14, 15, 15 and 26 have two modes, 14 and 15. Furthermore, it does not use all the scores. The mode of 14, 14, 16 and 99 is the same as the mode of 14, 14, 16 and 17.

The mode is of interest to people such as sellers of clothes. They need to know the sizes and colours of clothes that are the most popular. They do not buy the same number of each size of dress. They will buy more of the sizes that are most popular.

Topic 32 Displaying data on a pie chart

1 Introduction

The pie chart is so called because it looks like a pie. It is very effective for comparing quantities. Since the sum of the angles at the centre of the chart is 360°, the arithmetic is much easier if the total number of scores being illustrated involve factors and multiples of 360, for example 12, 15, 18, 20, 24, 30, 36, 40, 45 or 90. If there are 29 pupils in your class and you ask them a question about their favourite colour, include yourself to make a total of 30. If there are 31 in the class leave one out – perhaps a pupil who came late, or was talking!

2 Pie charts without calculating angles

Using children

Ask a group of pupils their favourite colour or team. (Limit the number of colours that they can choose from. Otherwise, there will be too many sectors on the pie chart.) Ask them to sit in a circle grouped according to their choices.

If in a class of 12, 3 liked green, 4 liked blue and 3 liked red, they would look like this.

This can be repeated for any size of class. The pupils have to sit in a circle, and be roughly equally spaced.

Other information can be illustrated using pupils.
For example, suppose pupils sleep for 8 hours, spend 6 hours at school, play for 3 hours, eat for 1 hour and help at home for 6 hours. 24 pupils can be used to show this. They sit in a circle. 8 pupils carry cards saying 'sleep', 6 carry cards saying 'school', and so on.

Using beads

Suppose there are 40 pupils in your class and you asked them how many brothers they have. 40 beads on a piece of string can be used to show the pie chart, with different types of beads being used for different numbers of brothers.

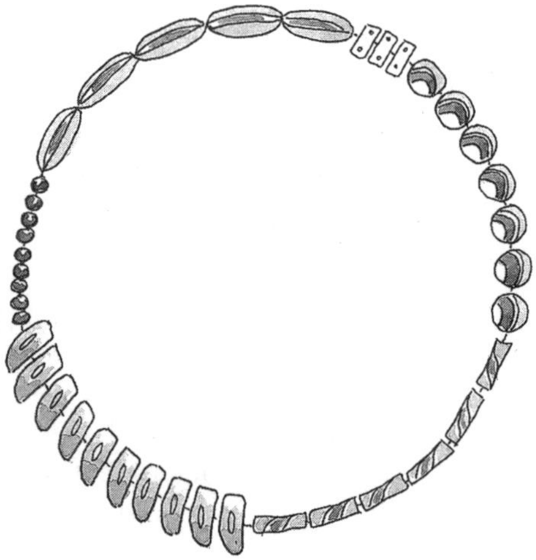

> **Activity**
>
> Prepare beads on string for the number of pupils in your class.
>
> Prepare circles on cardboard for your pupils to use as templates.

Folding a circle

A chart for 16 children can be made by repeatedly folding a thin paper circle. This circle has been folded into 16 sections.

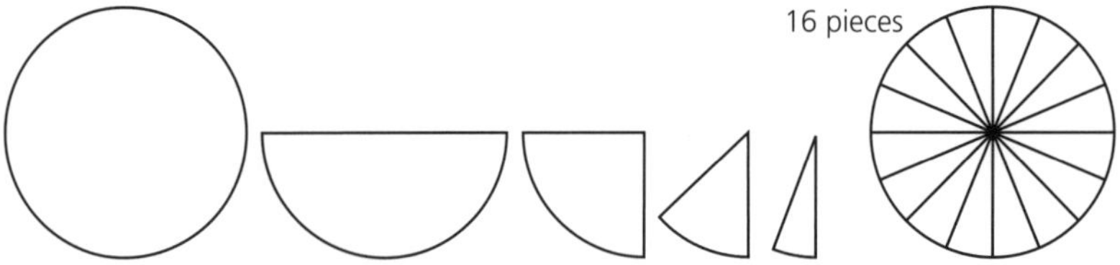

(The angle of each piece is 22.5°, $\frac{1}{16}$ of 360°.)

A pie chart to show 24 hours can be made by folding a circle into 2, then 3 then 2, then 2.

3 Calculating angles for pie charts

The pupils should by now appreciate that the size of the sector is proportional to the frequency being illustrated. Start with an example that is easy to calculate. Suppose that in a class of 12, one half, 6, liked green, 3 liked blue and 3 liked red. The pupils should be able to appreciate that one half of the chart should be for green and a quarter for each of the other colours.

The angle for each sector is:

Green: $\frac{6}{12} \times 360° = \frac{1}{2} \times 360° = 180°$

Blue or red: $\frac{3}{12} \times 360° = \frac{1}{4} \times 360° = 90°$

A more complicated example can then be given.

For example, in a class of 30, 11 pupils like football, 12 like netball and 7 like badminton. The angles for the pie chart are:

Football: $\frac{11}{30} \times 360° = 11 \times 12° = 132°$

Netball: $\frac{12}{30} \times 360° = 12 \times 12° = 144°$

Badminton: $\frac{7}{30} \times 360° = 7 \times 12° = 84°$

Note that the cancelling was done in the same way in each calculation, $360 \div 30 = 12$, which makes the calculations easier to do. It is also always a good idea that the pupils check that the angles calculated add up to 360°.

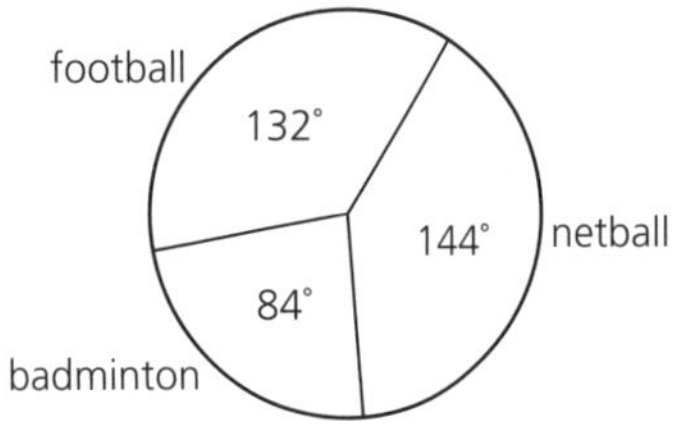

Activity

Collect economic and statistical data suitable for representation in pie charts.

4 Pie charts with percentages

If the information is given in percentages, the calculations are usually more complicated. A pie chart with percentages can be illustrated by putting 100 beads on a piece of string and arranging them in a circle equally spaced.

Section D Mathematics topics — measures

Topic 33 Money

1 Introduction

This is one of the most important and practical topics in the Primary school maths curriculum. Everyone, including children, needs to learn to use money intelligently and effectively. All countries now use a unit of currency that is divided into 100. This book will use dollars and cents, the most common currency around the world (symbol $). The word 'dollar' comes from the German word 'taler' which is short for 'Joachimstaler.' The taler coin was first made at the Joachimstal silver mine in the Czech Republic.

Obviously, teachers should use their local money in examples.
In many countries, the pupils will only have experience of large numbers when shopping. They will never use the coins and smaller unit of currency in their country.

2 Money-based activities

Pupils should do various practical activities. Shopping takes place in shops, markets, post offices and many other places. Pupils should have the opportunity to practise shopping. This cannot be done with real money but it can be done with toy money made from paper or card.

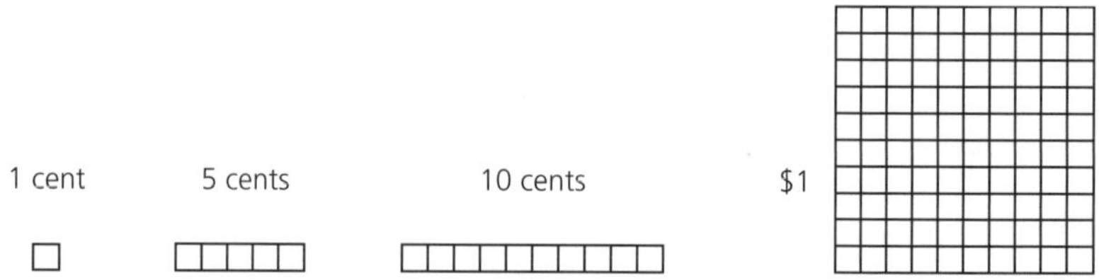

Cut out these shapes on squared paper. Pupils can then use them to pretend to shop.

Choose a suitable amount for your country that a small square should represent. For example, in a country such as Zambia where a large number of kwacha are needed to buy anything, each small square could be worth K1000. A 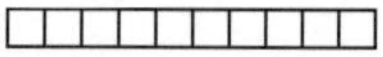 would then be worth K10 000.

Pictures of various items and their prices are needed. Pupils can then use their 'money' to go shopping. Some of the pupils will be shopkeepers and will have to calculate the change they should give their 'customers.' Customers should work out how much they spend and check that their change is correct.

This activity gives lots of practice with number operations. For example, if a packet of tea costs K4000, then three packets cost K4000 × 3 = K12 000.
Change from K20 000 is K20 000 – K12 000 = K8000.

A note on large numbers: In many countries, children have to become familiar with shopping with large amounts of the currency. It will help if the prices in the pretend shop are all kept to round figures and do not involve too much calculating with hundreds and thousands. Alternatively you could mark all the prices in US dollars and cents, British pounds and pence (£1.00 = 100 p) or another currency. This will keep the numbers small, and the pupils can enjoy pretending to be visiting somewhere else.

3 Comparing prices

Shopping

Where possible, encourage pupils to compare prices in different places. This is important, as shopping carefully can save a lot of money. You can organise pupils to compare prices in different shops near to the school. A shopping list will need to be drawn up and pupils should go to the various shops and find out prices. (You should arrange this first with the shop-owners, or the pupils should ask permission.) The prices in different stores can then be compared. The information can be used to answer many questions:

- Which shop is cheapest? Which is most expensive?
- What is the difference?
- Is the same shop cheapest for everything?
- What is the cheapest that the shopping bill can be?

Pupils can also compare prices of different sizes of packets of the same item. Usually (but not always) larger packets are cheaper (why?).

In the market

We buy much of our fresh fruit and vegetables from the market. We can compare the cost of items. Some fruit and vegetables will be cheaper towards the end of the day, as they do not keep overnight. The pupils can investigate this and prepare charts and tables.

> ***Activity***
>
> Investigate the cost of various items in local shops and markets. Use the results to draw the attention of pupils to different prices. Prepare suitable investigations for them to undertake.

The Post Office

The heavier a letter is, the more it costs to post. It also costs more to send letters abroad. Airmail is quicker but more expensive than surface mail. Find out details from a Post Office about the rates charged for different weights of letters and parcels.

4 Wages

There are three main ways of being paid for work:

- **Piecework** You are paid for work done, such as being paid for each dress made at home.
- **Salary** You are paid monthly. People such as nurses and teachers get paid in this way.
- **Daily paid** You are paid for each day's work. This is often the case for temporary workers.

Use rates of pay for various jobs in calculations. For example, if a man works a 40-hour week and is paid $80 per week, how much is that per hour?

5 Currency

Different countries use different currencies. Find out the conversion rates of various currencies against your currency from a bank or newspaper, and use them in calculations. For example:

> € is the symbol for 'euro', the currency used in much of Europe.

- $1 is worth €10, find the cost of the following in France.
  a book which costs $14 a car which costs $125 000
- How much would these cost in dollars?
  a pair of shoes which cost €125 a bicycle which costs €1085

6 Budgets

A useful activity for the pupils is the preparation of budgets. A budget could be prepared for a class party — and then the party could be organised and held. Pupils could prepare a household budget, finding out prices and working out what needs to be spent on groceries each week, and so on. All these activities involve the pupils in the collection of information, a very important skill.

Topic 34 Temperature

1 Introduction

Pupils need to learn to have a concept of different temperatures. They must also learn the vocabulary relating to temperature, and be able to measure using the standard units. They should develop an understanding of temperature before measuring. Before thermometers are introduced, pupils will need to know how to use words like 'hot', 'hotter', 'hottest', 'cold', 'warmer', and so on, and to be able to compare objects and the weather for heat.

2 Thermometers

A Swedish scientist Anders Celsius (1701–1744) devised a scale for measuring temperature in which the boiling point was 0° and the freezing point 100°. After his death, the scale was changed round and is now used everywhere. The freezing point of water at sea-level is 0°C and the boiling point of water at sea-level is 100°C.

The hotter an object is, the more agitated are the atoms in it. As the object cools, the atoms move more slowly. If it were possible to make them stationary, the temperature would be **absolute zero**, ⁻273.16°C. It cannot get any colder.

Pupils should have access to thermometers for work on temperature. Remind pupils to be careful with thermometers — they are not strong and break easily.

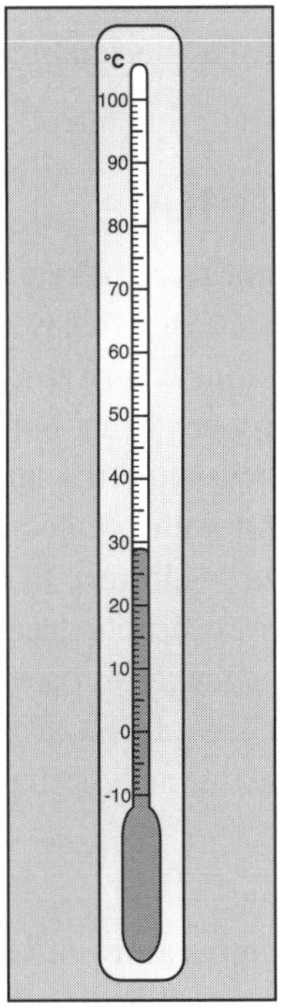

Room thermometer

This thermometer shows a temperature of 29°C, a hot day.

Class activities

Hang a room thermometer on the wall of the classroom and ask the pupils to record the temperature at the same time every day.

The class can then make a line graph that shows the variation of temperature at the same time of day throughout the week, term or year. They can make a note of the temperature on cloudy days, sunny days, hot days and cooler days. Here is an example of temperatures recorded at 11 a.m. every day for a week:

	Mon	Tues	Wed	Thurs	Fri
Temperature at 11 a.m. (°C)	28	27	30	31	28

The data can be shown on a graph:

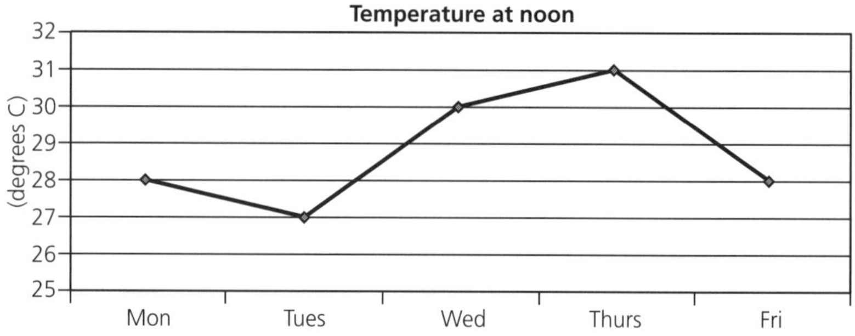

Clinical thermometer

Nurses use clinical thermometers to measure the temperature of the human body. The liquid inside a clinical thermometer is often mercury. This is poisonous, so a lot of care must be taken to ensure that the thermometer is not broken and the poisonous mercury released.

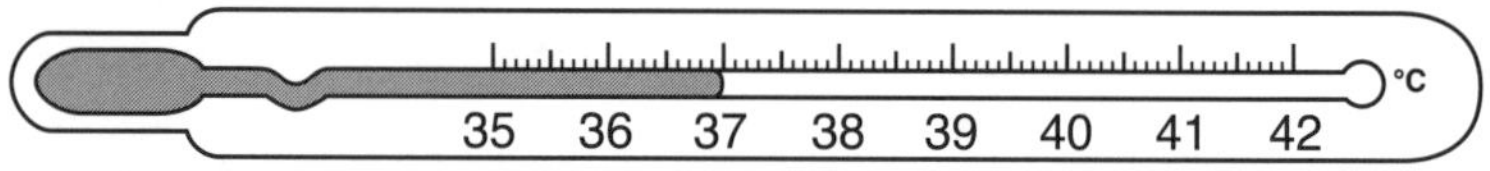

3 Low temperatures and high temperatures

Most pupils will have little experience of negative temperatures. Try to enable pupils to measure the temperature inside a fridge, inside the freezing compartment of a fridge, or inside a freezer.

Ask pupils to find out the temperature needed in an oven when baking a cake (most electric ovens have a thermometer). The temperature can reach 250°C.

4 Boiling water and altitude

Because water boils at a specific temperature, when food is boiled or steamed, it cooks at a constant temperature. We say water boils at 100°C, but this is only true at sea-level (or where the air pressure is equal to that at sea-level). If the water were being boiled high up a mountain, the air pressure would be lower and the boiling point of water would be reduced. (It is easier for water molecules to turn into vapour and boil away when the pressure is lower.) In many parts of the world water boils at a temperature several degrees less than 100°C. Consequently, here an egg takes longer to cook by boiling than at sea-level. The teacher should therefore not be surprised if the class measures the temperature of boiling water as less than 100°C.

5 Body temperature

A person who is well should have a temperature of about 36.9°C. If the body temperature rises by only a degree or two, the person will have a fever and feel ill. A thermometer is a very good and simple

Activity

Find the answers to these questions. Decide whether to ask your pupils to find the answers.

- Does water boil at the same temperature everywhere in the country?
- Could schools at different altitudes compare this?
- Are there high hills and low valleys near your school where a difference is noticeable?

way of checking the health of a patient. Because a small change in temperature of only a few degrees indicates that someone is sick, the thermometer has to be very accurate. Clinical thermometers have a scale running from about 36°C to 41°C.

The human body tries to stay at the same temperature whether the surrounding conditions are hot or cold. When the weather is hot, we sweat in order to reduce our temperature. When it is cold we shiver to raise the body temperature. There is a slight difference of temperature for different parts of the body — under the armpit the temperature is about 0.5°C lower than it is elsewhere. Other animals have their own normal body temperature, for example a chicken's normal temperature is about 42°C.

Topic 35 Mass and weight

1 Introduction

Pupils need to be able to estimate mass. They must learn the vocabulary relating to mass, and then learn to measure mass using the standard units. They should understand words such as 'heavy', 'light', 'heavier', 'lighter', etc. They should develop an ability to estimate mass before measuring.

2 Mass

Scientists distinguish between **mass** (the amount of matter in an object) and **weight** (the effect on the object of gravity). The mass of an object does not change; the weight changes when the effect of gravity changes, which it does from place to place (on the moon it has one-sixth of the effect it has on earth). In Primary school, we do not need to worry about this too much. Spring balances measure weight (because the object is being pulled down by gravity). Beam balances measure mass (because the objects on each side are being affected by the same force of gravity and we are comparing their relative amounts of matter). Both types of balance are calibrated with the same units, usually grams or kilograms.

3 The language of mass and weight

Heavy and light

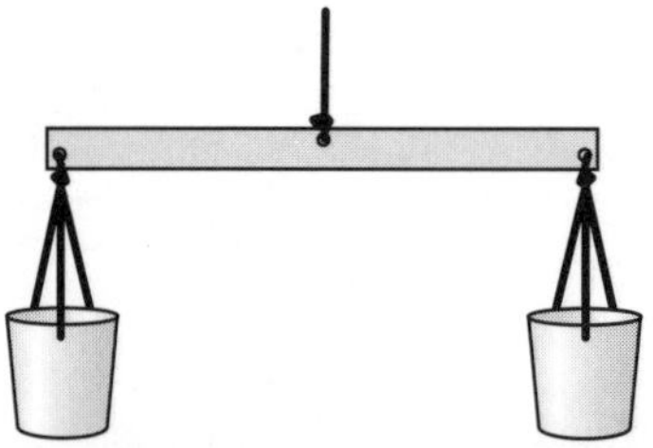

Give the pupils objects that are small but heavy, and large but light. Ask them to predict which are heavier (or lighter) by looking and then by holding them in their hands. Then ask them to compare the weights (masses) by using a home-made balance.

This leads to the realisation that on a balance, heavier objects pull the side that they are on down more.

4 Measuring mass

The standard measures of mass used are the gram (g), the kilogram (kg) and the tonne (t).

1000 cm^3 or 1000 ml or 1 litre of water weigh a kilogram.

To get an understanding of the different units, discuss the weight of a teenager (70 kg) and a young child (35 kg). An ear of rice, a fly and a drawing pin all have a mass of about 1 gram each.

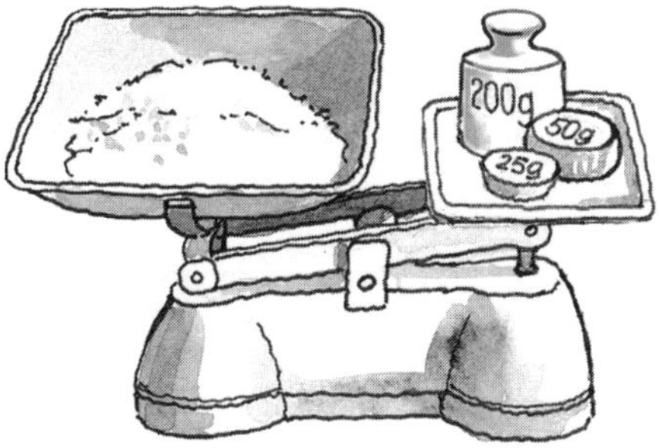

Pupils need practical experience of using scales and balances. They can weigh objects found in the classroom — some should be heavy and some should be light, so the pupils understand that different types of scales and balances are used in different situations. To find the weight of a drawing pin or a grain of rice the pupils will need to count and weigh a large number of them. A good number to use would be 1000. If 1000 drawing pins weigh 1 kg 55 g then each one has an average weight of 1.055 g.

Topic 36 Time

1 Introduction

Pupils need to learn that:

- **the length of time** of an event can be measured using the units of time, (seconds, minutes, hours, days, weeks, months and years).
- **any event can be fixed in time**, for example when saying, 'The netball match will be at 3 o'clock.' They have to learn to **tell the time**.

These two ideas are connected. The length of an event is the difference between the time at which it starts and the time at which it ends. For example, the length of a lesson that starts at 9.40 a.m. and finishes at 10.20 a.m. is the difference between the times 10.20 and 9.40, which is 40 minutes.

2 Preparatory work

Before pupils learn to tell the time they need to learn how to order events of the *past*, the *present* and the *future*. For example, they need to be able to put in order the following events: eating breakfast, going to school, getting up, washing.

Time is talked about everyday and children start learning about time at about 18 months, with words like 'early', 'late', 'now', 'later', 'before', 'after', 'then', 'soon', 'today', 'yesterday', 'tomorrow', and so on.

Tell the pupils stories and rhymes involving time.

3 Units of time

Pupils then need to learn about standard units of time — seconds, minutes, hours, days, weeks, and so on. The need to learn that a **day** is the time it takes the Earth to rotate once (the rotation being the cause of the sun moving across the sky). A **year** is the time it takes the Earth to go round the sun. A **month** is about the time it takes the moon to go round the Earth. The **day** is divided into 24 hours, an **hour** into 60 minutes and a **minute** into 60 seconds.

They need to develop a 'feel' for how long seconds and minutes are. This can be done with a **time guessing game**. For example, the pupils are required to sit quietly for 40 seconds, and are then told to write down how long they think they were silent. Those guessing between 35 and 45 seconds get 2 points, those guessing between 30 and 50 seconds get one point. Repeat the exercise for varying lengths of time (not too long) and the number of points of each pupil is added up.

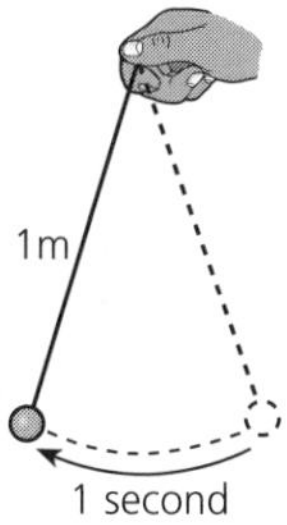

One second pendulum

A weighted string 1 m long will always swing for about 1 second. Pupils can make a simple pendulum clock by putting a weight on a piece of string 1 m long.

4 Telling the time

There are two ways of telling the time, 'digital' and 'analogue'. **Digital watches** show the time on a screen, as a number. They are popular because they are cheap and they are easier to use than watches with 'hands' (analogue watches). All you have to do is read the numbers. If a digital watch is showing 01:40:53, then it is 40 minutes and 53 seconds after one o'clock. All times are stated forwards. If the pupils have these watches, they will learn to tell time quickly. Their popularity and common use on electronic devices makes it harder to teach how to read a clock face.

However, pupils still have to learn how to read **clocks with hands** on them. For example, they will still have to learn that we say the time 07.45 as 'a quarter to 8'. They have to learn how to use a scale that is used in two ways at once. The shorter hand tells us the number of hours that have gone by, from 1 to 12. The longer hand tells us how many minutes, from 1 to 60, have gone by in that hour. To make things more difficult, times are stated in 'minutes after' and 'minutes before' the hour, and also use fractions. For example, '5.45' is also said as 'a quarter to six' or '15 minutes to six'. Charts with pictures and vocabulary are helpful when you are teaching this.

Stages in learning to tell the time

Teach the telling of time in stages, steadily getting more precise:

- in hours
- in half hours
- using quarter to and quarter past
- to the nearest five minutes
- to the nearest minute.

Start with a clock face that only has the hours on it.

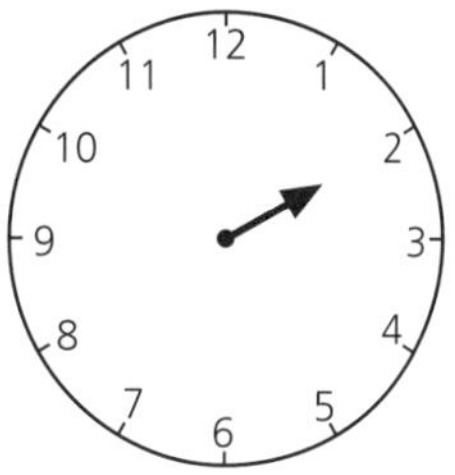

Then go on to clock faces showing the minutes.

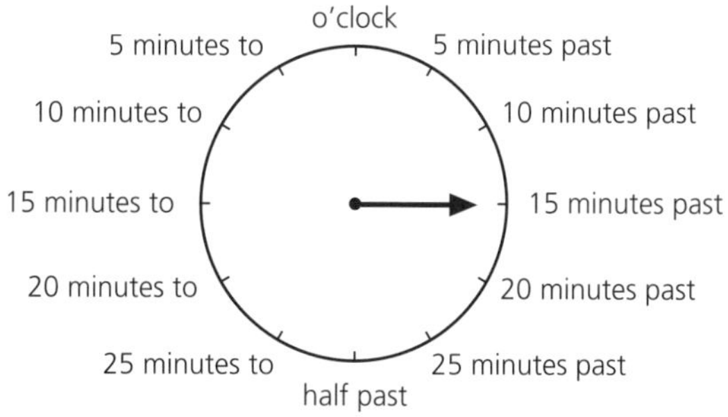

Then put them together.

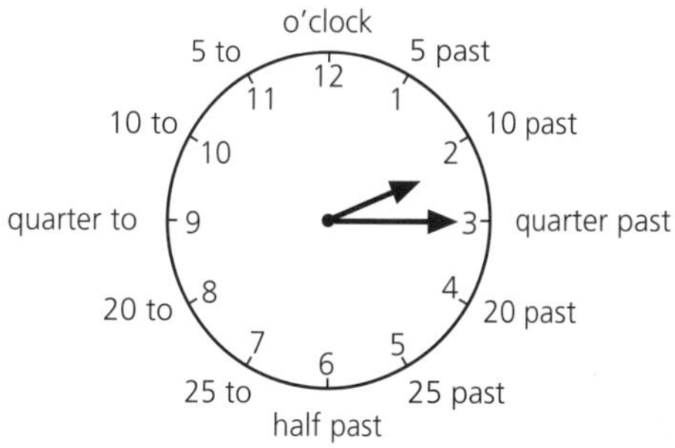

5 Times of the day

Link time to events during the day. A useful teaching aid is a chart that shows pictures of typical events at various times of the day — getting up at 6 a.m., in school at 11 a.m., and so on. Later on pupils can show the time in three ways for each picture, using a.m. and p.m., the 24-hour clock, and the usual way of talking about the time, as '$\frac{1}{4}$ to' and '$\frac{1}{2}$ past' the hour.

If you do not have a real clock, make a model clock with hands that can be moved. Pupils can be asked to read the clock and show lots of different times on the clock.

The children will have to learn the various ways of telling the time. There are 24 hours in a day, so the hour hand goes round twice in a day. If the time is 7.05, it is not clear whether this is 5 minutes past 7 in the morning or 5 minutes past 7 in the evening. For this reason, it is usual to say the time is '7.05 a.m.' or '7.05 p.m.'. '**a.m.**' is short for '**ante meridiem**,' which is Latin for 'before midday', and '**p.m.**' is short for '**post meridiem**' which means 'after midday'.

6 The 24-hour clock

This is used as another way of telling the difference between morning and afternoon or night. The time of 3 p.m. using the 24-hour clock is written as 15:00. For times after midday, 12 is added to the number of hours showing on the clock. **Bus, train and airline timetables** usually use the 24-hour clock because it removes the possibility of confusion. If timetables used the 12-hour clock there is a danger that some people would turn up for a plane at 7 o'clock in the evening when they should have been at the airport at 7 a.m.

7 Old clocks

Pupils can gain experience of time by using old-fashioned methods of measuring time.

Candles burn at a constant rate. Mark a candle at equal distances. The time to burn from one mark to the next is always the same and can be used to measure time.

Sand falls from a higher container to a lower container in a fixed time. You can make a sand clock to measure a time by taping two bottles together and putting sand in the top one. Measure how long it takes the sand to flow from one bottle to the other. Change the time the sand takes by changing the amount of sand or the size of the hole it falls through. (Sand timers are still used to measure the time needed to boil an egg.)

Shadow clocks (sundials) use the shadow made by a stick as the sun moves across the sky. The shadow moves and the movement is used to tell the time.

Activity

- Put a stick in the ground in a sunny place and mark where the shadow is each hour.

  Now you can use the marks as a clock.
- Note the positions of the shadow at different times of the year and mark them. The hour markers will not be the same distance apart.

Sundials have two disadvantages. They do not work when it is dark or cloudy and the sun does not move through the sky at the same speed throughout the year.

8 Timetables

These should not to be confused with 'times tables' in multiplication! Airline timetables show the departure and arrival times of planes. Each airline has a two-letter code of its own; each of its flights has a number. Get examples of timetables for airlines, railways and buses, display them in the classroom and teach the pupils how to use them.

Thirty days has September,
April, June and November.
All the rest have thirty-one,
Except February alone,
Which has twenty-eight days clear,
And twenty-nine in each leap year.

9 The calendar

Make a chart with a calendar for a whole year giving the number of days in the months and showing important days — end of term, public holidays, and so on. Members of the class can fill this in. The children will need to learn the names and lengths of the months. Remembering the lengths of the months is not easy, as the system is not very logical.

10 A time line

The pupils can prepare a time line for the class giving lots of dates in recent history and in their lives. This can make a most attractive addition to the classroom. You can make it by joining pieces of paper together with sticky tape. For example,

Put events like the Year of Independence, Olympic years, the birth years of important people, years when important world events happened, etc. on the chart.

Here is a simple time line for a school day.

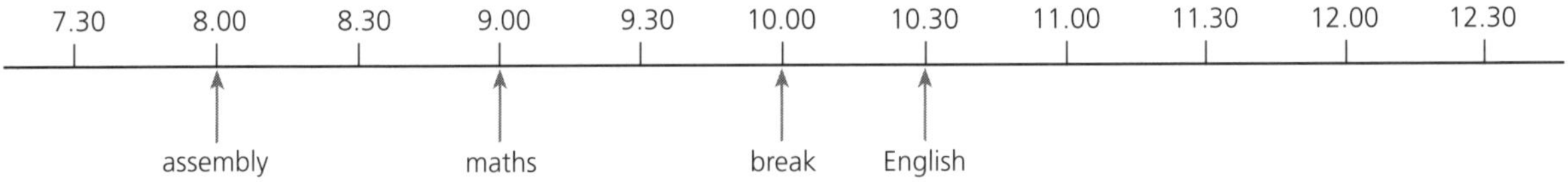

11 Adding and subtracting times

Pupils can become confused because the addition and subtraction of times does not seem to follow the ordinary rules of arithmetic. 15 minutes after 6.50 is 7.05, and yet 650 + 15 = 665. Why is this? The pupils have to remember that there are *60* minutes in an hour. (Time works in base 60, not base 10.) Encourage them to think about the minutes and hours separately, then they can see that at 6.50 there are ten more minutes until 7 o'clock, and a further 5 minutes to add on. 5 hours after 8 a.m. is 1 p.m., yet 5 + 8 = 13. (This makes sense if you think in the 24-hour clock system.) 8 hours after 20:00 is 04:00. (Again, think in '24 hours', base 24.)

Calculations of journey times have to be handled carefully. For example, if a plane departs at 16:45 and arrives at 19:40, how long is the flight?

There are two ways of doing this.

- **Find the difference between the two times**

$$\begin{array}{r} \overset{18}{\cancel{1}\cancel{9}}:\overset{60+}{40} \\ -\ 16:45 \\ \hline 2.55 \\ \hline \end{array}$$

You cannot subtract 45 minutes from 40 minutes, so you need to decompose one hour into 60 minutes.

- **Count forward**

  16:45 to 17:00 is 15 minutes
  17:00 to 19:40 is 2 hours 40 minutes
  The total time is 2 hr 40 min + 15 min = 2 hr 55 min.

The pupils will need plenty of practice at these sorts of calculations.

Changing hours to minutes and vice versa uses the fact that 60 minutes = 1 hour.
For example, 5 hours 40 minutes = 5 × 60 + 40 = 340 minutes

$$8200 \text{ minutes} = \frac{8200}{60} = \frac{410}{3} = 136\tfrac{2}{3} \text{ hours}$$

Activity

- Investigate the history of the calendar and the origins of the names of the days and months.
- Investigate why the day is divided into 24 hours, the hour into 60 minutes, and so on.
- Consider which method of calculating time you prefer, and if you prefer different methods depending on the problem to be solved.

Activity

A time graph

Members of the class could start growing plants at the start of term, and regularly measure their heights week by week through the term. The height of each plant makes a simple bar graph or time graph.

The bars could be replaced by pictures, drawn by the children, of plants of the correct height.

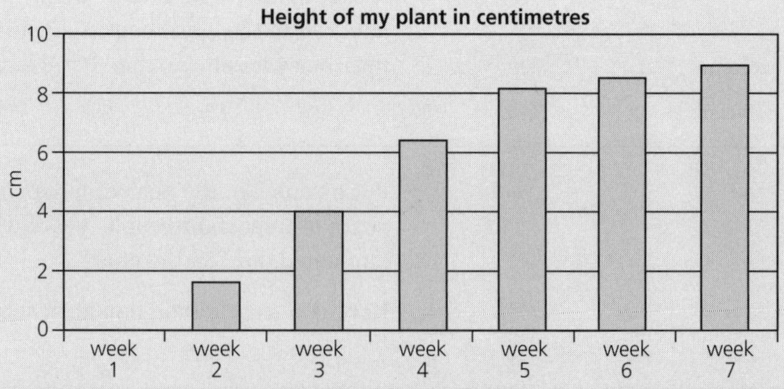

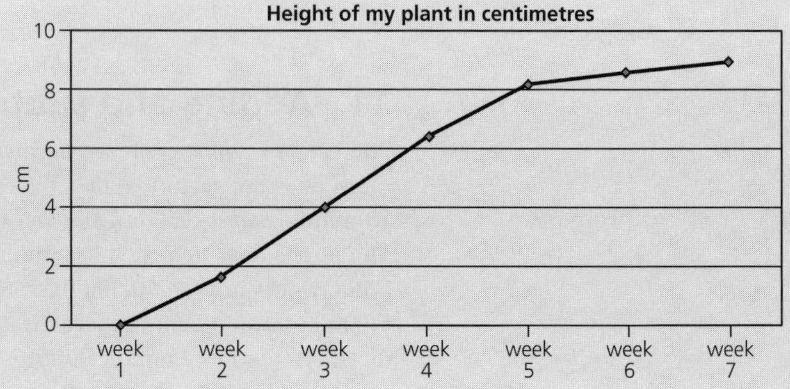

Section E Mathematics topics — shape and space

Topic 37 Angles

1 Introduction

Angles are very important in school geometry. They can be thought of in two ways:

- as an amount of turning
- as a sharp point or corner.

An angle as an amount of turning

Introduce angles as turning by asking pupils to stand and **turn** to their left and right (**anti-clockwise** and **clockwise**). The larger the turn, the larger the angle. There are many other examples of turning that pupils can give:

- clock and watch hands (talk about clockwise and anti-clockwise)
- handles, doors and windows
- opening a box lid
- scissors
- putting in a screw, or tightening a nut.

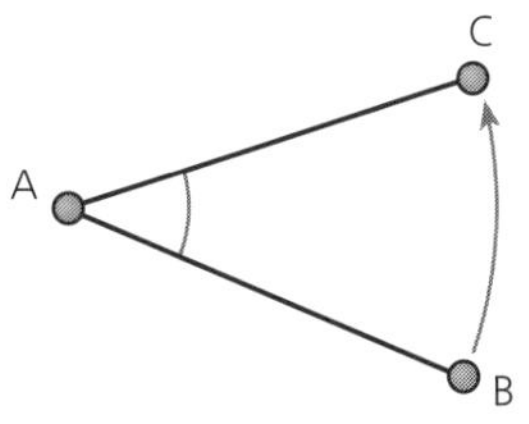

Three pupils holding a piece of string can illustrate this well. Pupil A holds the middle of the string. Pupils B and C stand together, each holding one end. C starts to walk round, keeping the string tight. The string makes $\angle BAC$, which gets steadily larger.

An angle as a sharp point or corner

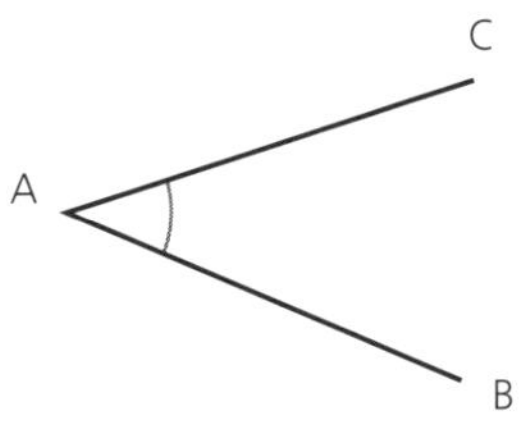

This can also be thought of as the opening between two rays or arms. Look for pointed objects in the classroom, such as pencils. The smaller the angle the sharper the point. **An angle** has two **arms** and a **vertex** (pl: vertices). In the diagram, the vertex is A, the arms are **line segments** AB and AC and the angle is written as $\angle BAC$ or $\angle CAB$ (or $B\hat{A}C$ or $C\hat{A}B$). AB can also be thought of as turning to AC.

Naming angles

Pupils need to learn how to name angles, to be able to refer to angles marked on figures, and to join up points to form named angles.

Mark a point on the board and say, '*This is a point.*' (Do not try to *define* a point.) Mark two points on the board, call them A and B, and join them with a straight line and say; '*This is straight line AB.*' Folds in paper or holding taut string give simple examples of lines. Draw another line (CD) to cross the first (at Z). Show the pupils the angles that are formed, and name them ($\angle AZC$, $\angle CZB$, etc.)

2 Special angles

The right angle or quarter turn

A complete turn is the basic unit for measuring angles. The quarter turn or right angle is probably the most important angle. It can be introduced as a **left turn** or **right turn**. Pupils can make a right angle from any piece of paper.

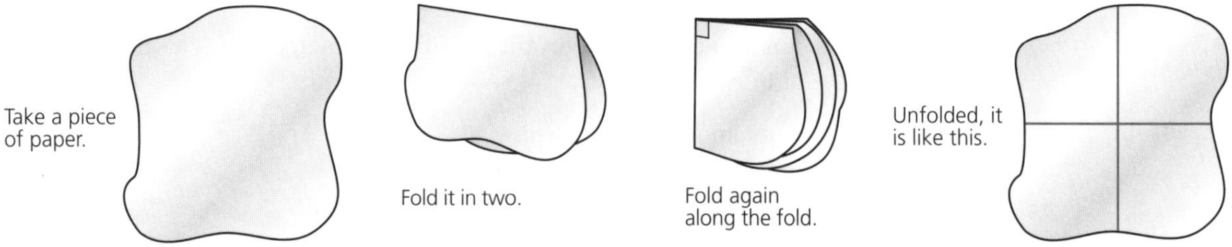

The pupils can use the right angles they make to find other right angles in the classroom — on windows, books, furniture, walls, etc. They are everywhere. They can also compare other angles to see if they are larger or smaller than right angles.

Straight angle

Two right angles put together make a **straight angle**. To a child, a straight line is not obviously an angle. It can be seen as an angle by using joined sticks.

Acute, obtuse and reflex angles

Pupils can now learn about **acute**, **obtuse** and **reflex** angles. Acute angles are smaller and sharper than right angles, obtuse angles are larger and blunter than right angles, reflex angles are larger than straight angles.

These angles can be illustrated by the use of string as described earlier. When B and C are close together the angle is acute. As C moves round, the angle becomes a right angle, an obtuse angle, a straight angle, and then a reflex angle.

Pupils can make **angle discs**. Give each pupil two small congruent circles. Ask them to colour one and cut a radius on both of them. Join them so that they fit together as one circle. They can be turned to show different size angles.

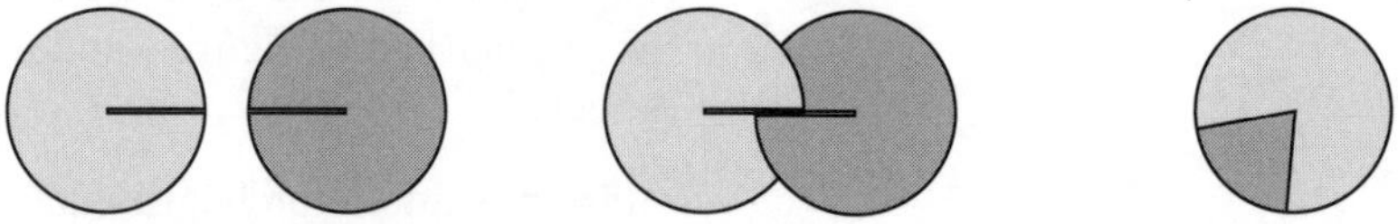

Children can **compare** angles without measuring them. For example, a door can be opened a little or it can be wide open.

3 Measuring angles

Using a **protractor** to measure angles in degrees is quite difficult. The pupils can start by using a larger unit. Give each pupil a semi-circle of paper. They can fold it in three and then into two making six parts. This is now a simple protractor and can be used to measure angles in 'parts' or to the nearest number of parts.

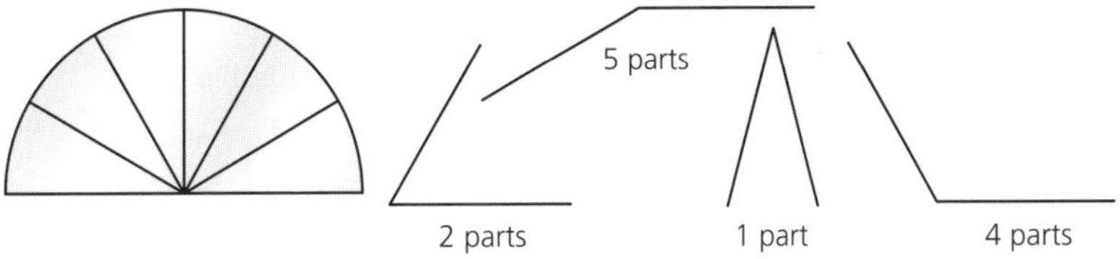

Using a protractor

Angles are measured in degrees. There are 360 degrees (360°) in a complete turn. (The ancient Babylonians divided the circle up into 360 parts.)

So a right angle is one quarter of 360° = 90°,
a straight angle is one half of 360° = 180°.

When measuring an angle, put the centre of the protractor on the vertex of the angle, 0° on one arm, and read the size of angle.

The protractor is quite tricky to use. There are two scales, going in opposite directions. The vertex and one arm of the angle have to coincide with the centre and base line of the protractor, which are not at the edge of the protractor. (The teacher's board protractor does not have this second problem, as it has no edge.)

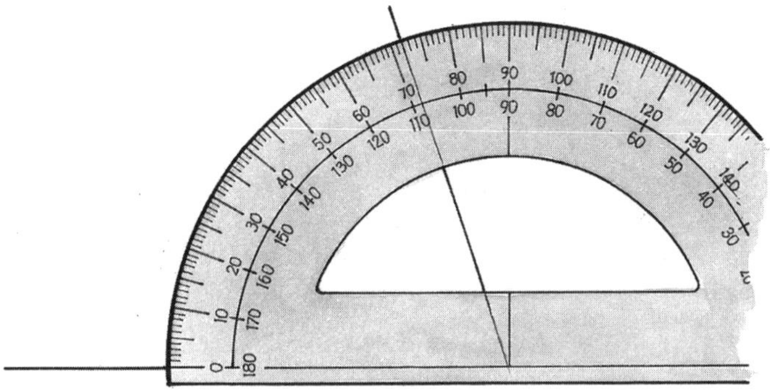

Estimating angles game (2–6 players)

Since there are two scales on a protractor, it is important that pupils learn to estimate angles as a quick check that they have read the correct scale. It is common for children to read from the wrong scale, so an angle of 27° may be stated as an angle of 153°. A game can help pupils learn to estimate.

1 Each pupil draws an angle.
2 They estimate the size of their angle, and of the other angles, and write down their estimates.
3 They measure one angle (not their own).
4 They find the difference between the actual measurement and their estimate. A point is given to the person who has the best estimate.

4 Common problems pupils have with angles

- Long-sided angles are often thought to be bigger than short-sided angles.

But  is smaller than 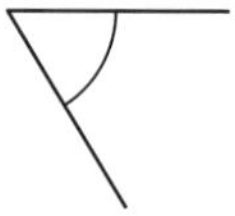 because there is less turning.

- Different looking angles are often thought to be different sizes. For example these angles are the same although they are at different orientations.

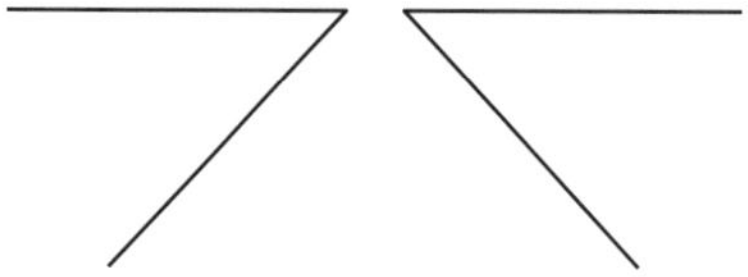

- The centre of the protractor is not placed at the vertex.

A 'one-way' protractor with no edge is a useful tool as a start to learning to use an ordinary protractor.

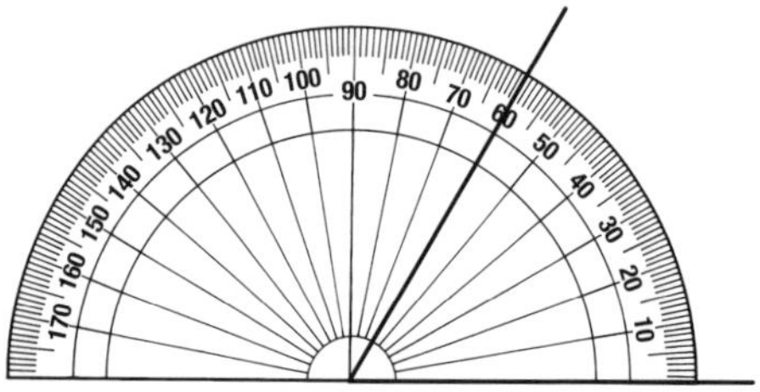

- Degree markings on a protractor are very small. Start by asking pupils to measure angles which are a multiple of 10°, then angles that are multiples of 5°, and then other angles.

Activity

Make one-way protractors for use in class.

Draw circles on card using a round tin. Cut them into semi-circles. Mark angles every 10°.

5 Drawing angles

Pupils can now learn to draw angles of various sizes. Once again they should start with angles that are a multiple of 10°, then angles that are multiples of 5°, and then other angles.

Topic 38 Angles at a point and on parallel lines

1 Introduction

This topic looks at the properties of angles associated with lines that meet at a point and lines that are parallel — **vertically opposite, alternate, corresponding** and other angles. The pupils can discover many of these results for themselves by measuring.

2 Parallel lines

If you just say that parallel lines are lines that never meet, a pupil can give lots of examples of lines that do not meet. For example, a line on the front cover of this book and a line on the wall. What is missing is *in the same plane.*

Parallel lines are lines *in the same plane* that never meet.

The lines are the same distance apart everywhere. They never meet, even when extended. They are like railway lines.

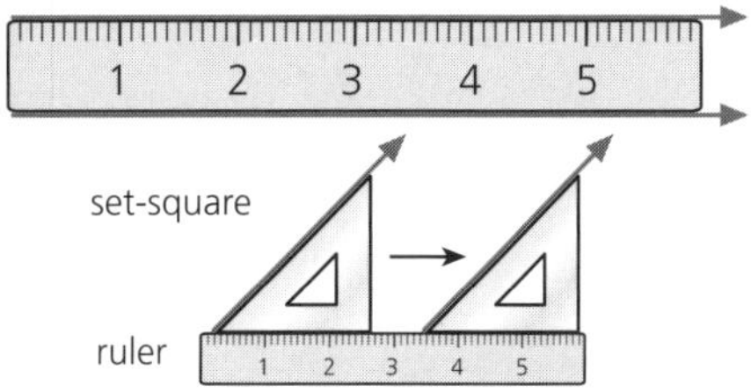

Drawing parallel lines

There are several ways to draw parallel lines. Three ways are:

- using the two sides of a ruler
- using a set square and a ruler
- using the lines in an exercise book.

(The pupils need to be told the use of arrows to indicate two lines are parallel.)

3 Special angles

Angles at a point

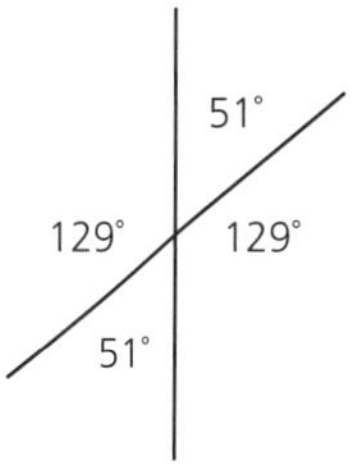

Pupils should draw two lines that cut (intersect) each other. Then they should measure all the angles.
What do they notice?
They have two pairs of equal angles.
These are called **vertically opposite angles**.
(*Vertically* not because they are vertical but because they are on the same vertex. *Opposite* because they are on opposite sides.)

Angles on a straight line

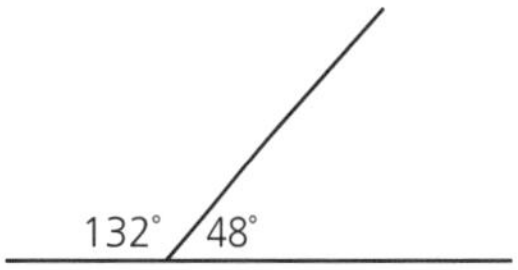

Pupils should draw a straight line and one line to meet it. Then they should measure the two angles formed.
What do they notice?
The angles add up to 180° (not so easy to notice as the previous result).

Supplementary and complementary angles

Two angles that add up to 180° are called **supplementary angles**. The three angles 50°, 60° and 70° also add up to 180° but they are not called supplementary because there are three of them.

Two angles which add up to 90° are called **complementary angles**. Three angles which add up to 90° are not called complementary.

Complementary and *supplementary* are very similar words and easy to confuse. One way to distinguish them is to note that *c* comes before *s* and complementary angles are smaller than supplementary ones.

Alternate, corresponding, and co-interior angles

The angle properties of parallel lines can be discovered by measurement.

Pupils should draw two parallel lines and a **transversal**. Then they should measure the two marked angles. What do they notice?

- The angles are the same.
- They are called **corresponding angles** as the places they are in, with relation to the transversal, are the same.
- They are sometimes called **F** angles.

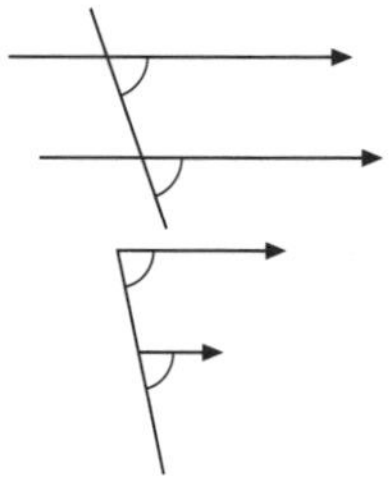

The pupils should measure these marked angles. What do they notice?

- These angles are also equal.
- They are called **alternate** angles.
- They are sometimes called **Z** angles.

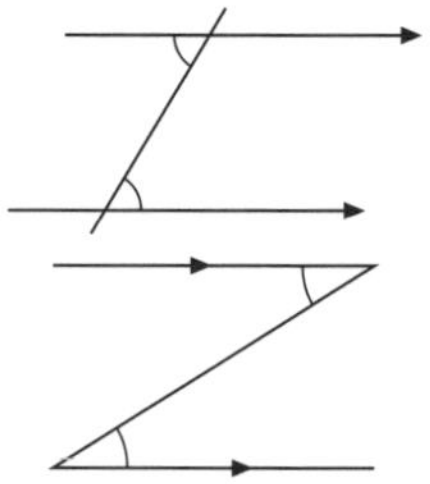

Finally, the pupils should measure these marked angles. What do they notice?

- These two angles have a sum of 180°.
- They are called **co-interior** or **allied angles**.
- They are sometimes called **C** angles.

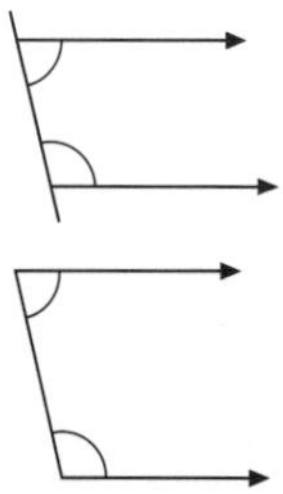

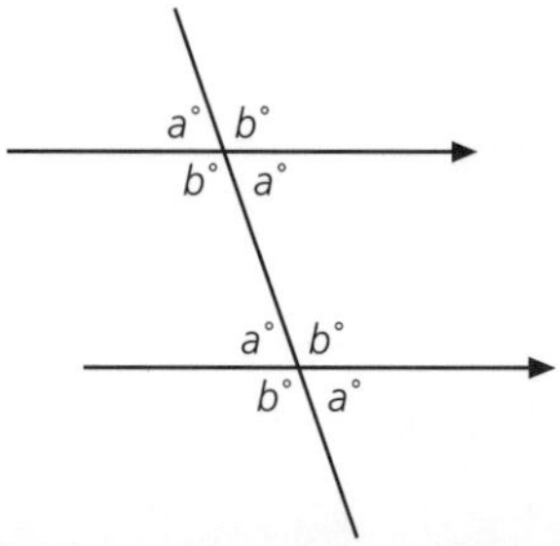

All these results can be shown in one diagram.
All the angles marked $a°$ are equal.
All the angles marked $b°$ are equal.

$$a° + b° = 180°$$

Topic 39 Angles of triangles

1 Introduction

There are many angle properties that pupils have to learn. They can find these out for themselves by measurement and other ways.

2 The sum of the angles is 180°

When you teach this, you can help the pupils to find out for themselves that the angle-sum of a triangle is 180° by doing one or more of these activities.

- **By measurement**
  - (i) Each pupil draws three large triangles in his/her exercise book. At least one triangle should be obtuse angled.
  - (ii) Each pupil measures all the angles of the three triangles.
  - (iii) Each pupil finds the sum of the angles of each of his/her triangles.

  You can make a table of the results on the board like this:

Name	**Angle**	**Angle**	**Angle**	**Sum of angles**

> **Note:** If pupils' answers are very different from 180°, it means that they have measured or added up incorrectly. Some pupils will get answers of 179° or 181°. These answers may be due to rounding.
> For example, suppose the angles were 60.4°, 70.4° and 49.2° to the nearest tenth of a degree. Measured correctly, to the nearest degree, these angles would be 60°, 70° and 49°.
> $60.4° + 70.4° + 49.2° = 180°$ but $60° + 70° + 49° = 179°$.

Ask the pupils what they notice. They should notice that the three angles of any triangle add up to 180°.

- **By cutting**

(i) Each pupil cuts out a triangle from rough paper and marks the three angles.

(ii) They tear off the corners of their triangles and put them together as in the diagram.

Ask the pupils: 'What does this tell you about the angles marked *a*, *b* and *c*?'
(Answer: 'They form a straight line.')
'What does this tell you about the sum of the angles marked *a*, *b*, and *c*?'
(Answer: 'Their sum is 180°.')

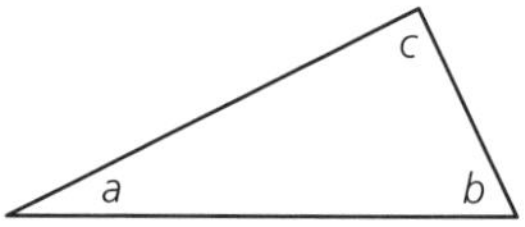

- Tesselations of triangles
  (i) Give each pupil a labelled, cardboard triangle.
  (ii) The pupils should draw around their triangle and make a tessellation like this.

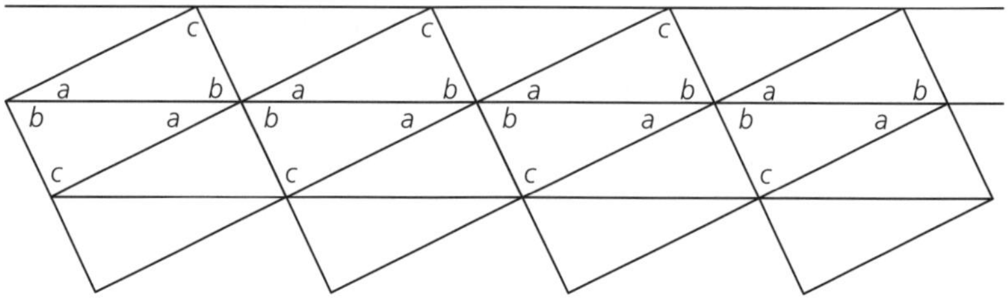

(iii) Label the angles *a*, *b*, and *c*.
Ask the pupils what they notice about *a*, *b*, and *c*.
They make a straight line and add up to 180°.

- **Walking around a triangle**

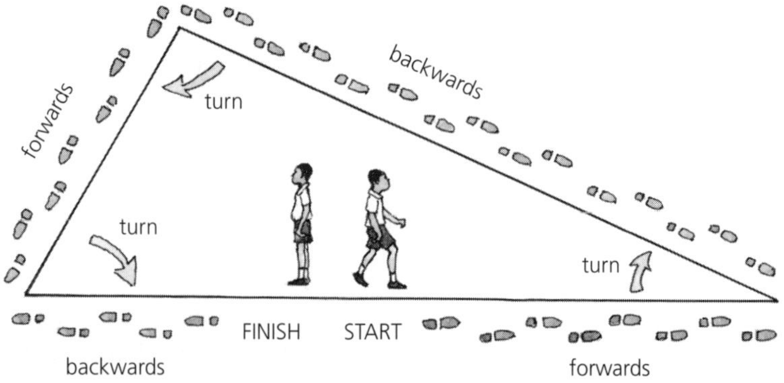

Draw a large triangle on the ground or make a triangle out of pupils. A pupil walks around the triangle turning through each angle of the triangle at the vertices. Ask the pupils to describe what has happened to the pupil. He/she will have turned around (be facing the other way) when he/she gets back to the starting point. This shows that the angle-sum of a triangle is a half-turn or 180°.

Activity

- What do the children need to know before they can learn this topic?
- Try out these activities to show the angle-sum of a triangle is 180°.
- Which of them do you prefer? Why?

3 Special triangles

Many triangles are **scalene** (their sides are all different and their angles are all different). Some triangles have special names.

Right-angled triangle

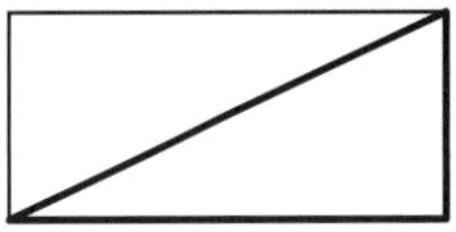

Cut a rectangle in two along a diagonal.
These are right-angled triangles.

Isosceles triangle

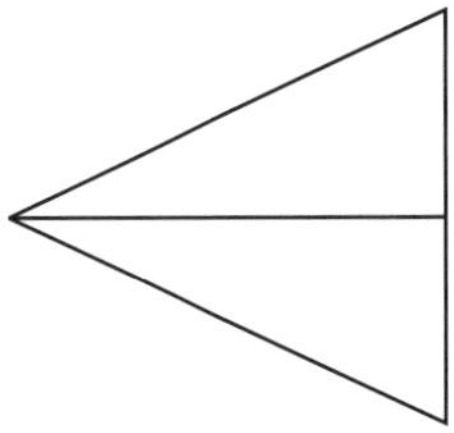

Take the two right-angled triangles you have made and put them together. (Stick them together with tape.) The shape is an isosceles (a very difficult word for many pupils) triangle.

Which line is the axis of symmetry?

Equilateral triangle

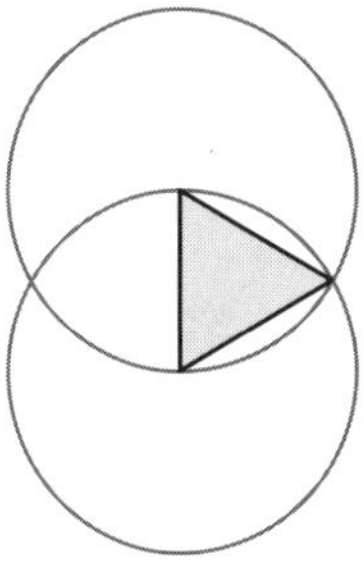

Draw a circle with a pair of compasses. Mark the centre of the circle.
Mark a point on the circle. Draw a circle from there with the same radius.
The two centres and a point where the circles intersect make an equilateral triangle.

> ***Activity***
>
> - Make an isosceles triangle from two right-angled triangles. Use it to demonstrate the properties of isosceles triangles. (Two sides equal, two angles equal)
> - Make an equilateral triangle. Use it to demonstrate the properties of equilateral triangles. (All sides equal, all angles equal)
> - How many lines or axes of symmetry do the triangles have?

4 Exterior angles of a triangle

Ask pupils to draw this shape on a piece of paper and measure the angles marked *a, b* and *f*. Then measure the angles marked *b, c* and *d*. Finally, measure the angles marked *c, a* and *e*. Ask them *'What do you notice?'*

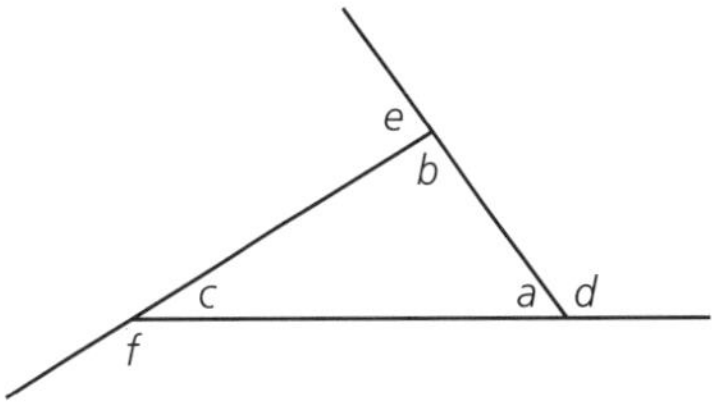

There is a detailed discussion of symmetry in Secton E, Topic 49.

> ***Activity***
>
> How can you help pupils solve common problems like these about the angles of triangles?
>
> - Two angles are given and pupils are asked to find the third angle. What is the most effective way of working this out?
> - When one angle is a right angle, what is the sum of the other two? How will you help the pupils to discover that the sum of the other two angles is 90°?
> - In an isosceles triangle, one angle is 70°. What are the other two? (There are two possible answers.)

Topic 40 Bearings

1 Introduction

Pupils need to learn how to describe directions. They start with north, south, east, and west. This can be done when they learn about right angles.

East is where the sun rises.
West is where the sun sets.

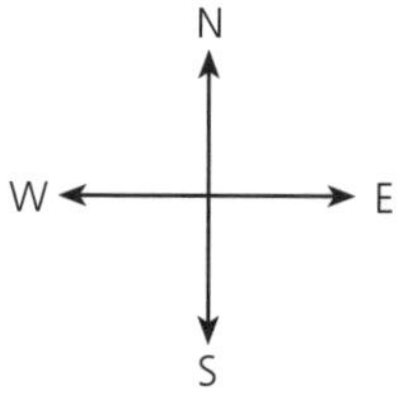

Stand with your right side facing east and left side facing west, and you are looking north. Turn around and you are facing south.

Sometimes we need more precise directions than these, four more directions, northwest, northeast, southwest and southeast are introduced halfway between N, S, E and W.

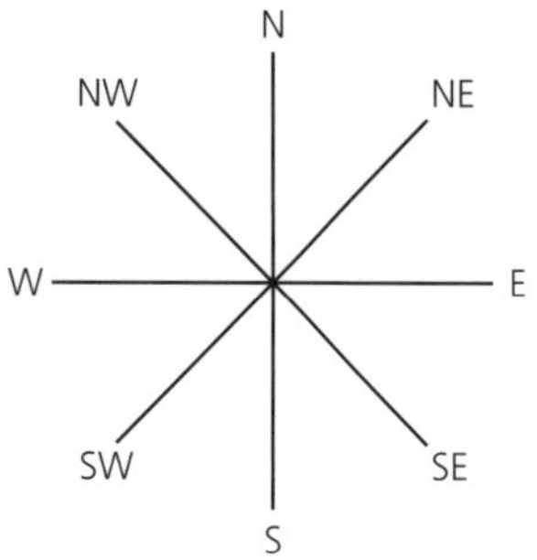

A practical illustration

These directions can be illustrated by five pupils, each given a label with N, S, E, W or O on them. Ask them to stand in a circle with O at the centre. Add four more pupils with NW, SW, SE and NE cards. Pupil O can then turn and face in different directions.

2 Three-figure bearings

Pilots and sailors need more accurate directions than these and three-figure bearings are used. The direction is found by starting from north and turning clockwise. Everyone starts from north because that is the direction magnetic compasses point. Everyone has agreed to turn clockwise when talking about bearings. If they had not, you would have to say 'clockwise' or 'anti-clockwise' each time. The bearing is given as three-figures because when people are listening to a pilot talking on radio, they need to be certain that they have heard his bearing correctly. If they only hear two figures they know the bearing is incomplete.

A bearing of NE would be given as 045°. To go due east a pilot would be given a bearing of 090°. Similarly a bearing of 5° would be stated as 005°.

3 Teaching bearings

Emphasise that a quarter turn form N to E is 90° and a half turn from N to S is a 180° turn. Then consider other angles such as 60°.

For example: A is SE of B. How far do I turn from north to face him?'

Answer: 135°. We say, 'The bearing of B from A is 135°.'

Give more examples, using NW and so on, and other angles.

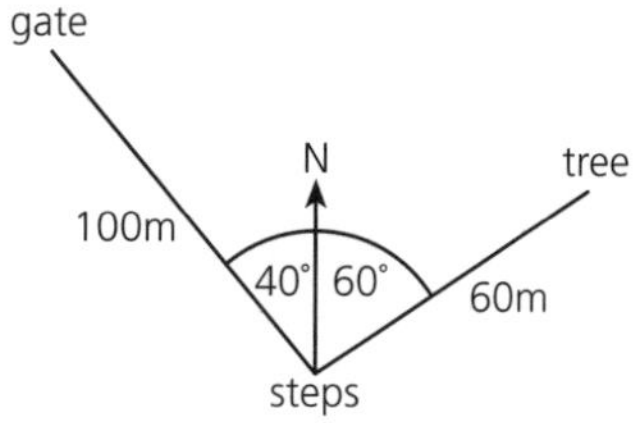

Compasses in school grounds

Stand in the centre of the school yard. Find north with a compass and mark the direction on the ground. Find east, west and south by using angles and mark those directions on the ground too. Find the bearings of other places you can see. Use the information to draw a plan of the school. To draw an accurate plan you will also need to measure the distances to these places.

4 The bearing of A from B, and B from A

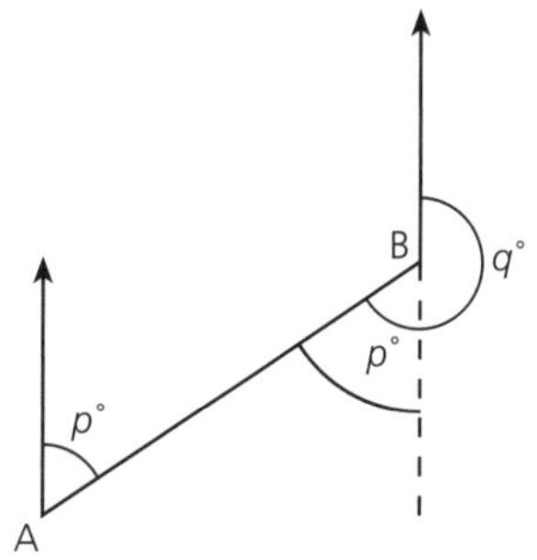

The relationship between these is quite easy to express.
If the bearing of B from A is $p°$, and the bearing of A from B is $q°$, then from the diagram, $\boldsymbol{q° = p° + 180°}$.

This means that if the bearing of B from A is 075°, then the bearing of A from B is found by adding 180°, 075° + 180° = 255°. If the bearing of P from Q is 320°, then adding 180° would give a bearing of Q from P of 320° + 180° = 500°. This is more than 360° so subtract 360° giving 500° – 360° = 140°. (Alternatively subtract 180° and the bearing is 320° – 180° = 140°, since 180° – 360° = $^{-}$180°.)

Topic 41 Quadrilaterals

1 Introduction

This topic looks at the angle and symmetry properties of quadrilaterals. (Area formulae are in Section E, Topics 43 and 44.) For each shape, pupils should be involved in practical work. The work should go through several stages:

- Show the pupils examples of different quadrilaterals. These could be cut out from paper, or be things found in the classroom or brought to the class.
- Discuss the name and the definition of each quadrilateral with the pupils.
- Develop the discovery of the properties of each quadrilateral in a practical way.

2 The sum of the angles of a quadrilateral

The angle-sum of a quadrilateral is 360°. This can be shown to the pupils in at least three ways.

- **By measurement**
  Draw two quadrilaterals on the board, one small and one large. Ask the pupils, '*Which is larger?*' They pick the large one. '*Which has the larger angle-sum? How can we find out?*' Pupils suggest measuring the angles. Tell the pupils to draw three quadrilaterals in their exercise books, one big, one middle-size and one small (but not too small). They then measure the angles, add them up and find that their sum is about 360°, no matter what size the quadrilateral is.

> **Note on accuracy:** There will be inaccuracies due to measuring, in the same way as there were for triangles. Some pupils will find totals of 359° and some 361°.
> For example, if the angles were 100.6°, 100.6°, 100.6° and 58.2°, with a sum of 360°, these would be measured as 101°, 101°, 101° and 58°, with a sum of 361°. The difference of one degree is not the pupil's fault.

- **By deduction from the triangle angle-sum**
  Pupils know that the sum of the angles of a triangle is 180°. Any quadrilateral can be divided into two triangles. Therefore the angle sum is $2 \times 180° = 360°$.

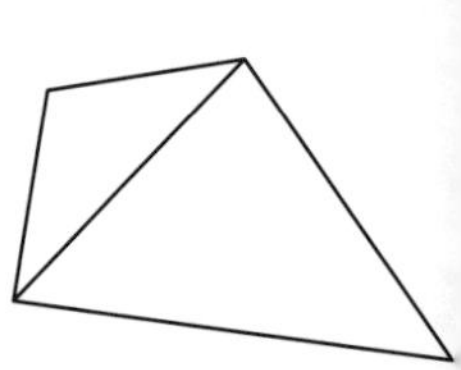

- **Walking around a quadrilateral**
  Make a quadrilateral out of pupils and ask a pupil walk around the shape. (This was done with a triangle in Section E, Topic 39.)

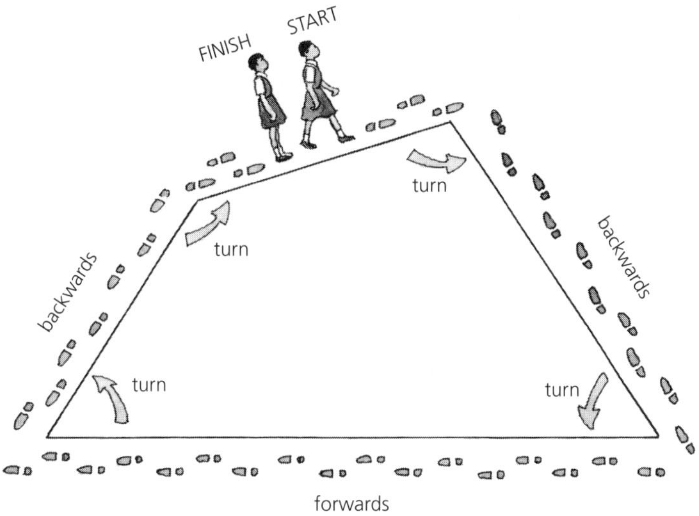

The pupil will turn around completely, and will be facing the way he/she was at the beginning. The angles of a quadrilateral make a complete turn of 360°.

3 Rectangle

This is a very common shape in the classroom and in people's houses (though not in nature).

Give each pupil a paper rectangle. (You can make many small rectangles by folding and cutting one piece of paper.) Ask the pupils: '*What can you tell me about it? Where can you see this shape?*'

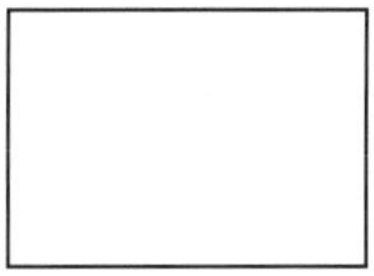

- It has four straight sides at right-angles to each other.
- All the angles are 90°.
- Opposite sides are the same length.
- There are two pairs of equal length sides.
- Books, bricks, photos, etc. all show examples of rectangles.

There is a full discussion of symmetry in Section E, Topic 49.

Ask them if they know what it is called (a **rectangle**). To help them remember the word, make the connection between 'rectangle' and 'right angles.'

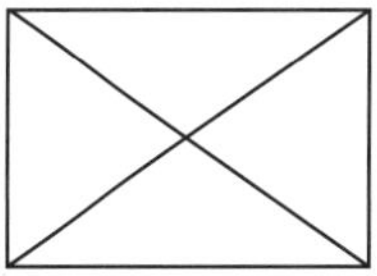

Tell the pupils what is meant by 'the diagonal' and ask them what they can say about the diagonals. (They are equal and cut each other in two.)

Look at the line (or folding) symmetry of a rectangle. You can ask: '*How many different ways can you cut a rectangle in two equal parts that fold onto each other?*' Some pupils will say four, (including the diagonals). Show them by folding that a diagonal is not a line of symmetry.

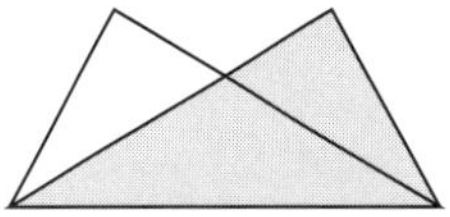

> *A question for pupils*
>
> Why are writing paper, books, etc. usually rectangular?

For rotational symmetry ask: *'If it is turned around once, how many times will it fit into its outline?'* (Two)

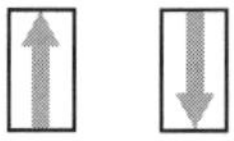

4 Square

Pupils can make a square from a rectangle. They fold the short side onto the long side and cut or tear the extra bit off.

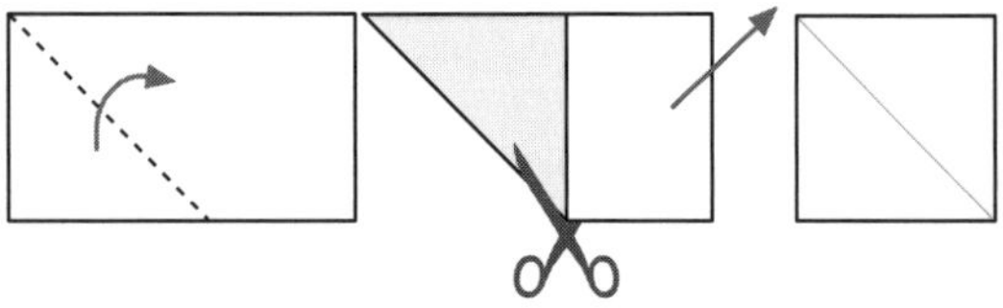

Ask them what they can tell you about the shape.

- The sides are the same lengths.
- All the angles are 90°.

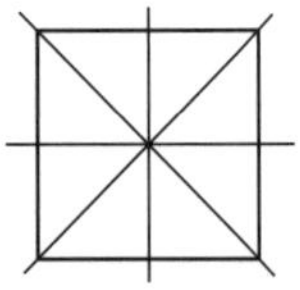

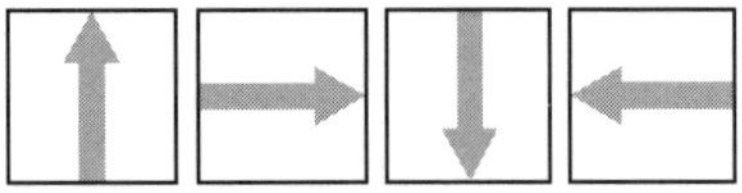

Ask if they know what it is called (a **square**). Ask the pupils to draw the diagonals and they will be able to tell you that they cut each other in two at right angles.
When discussing line symmetry, they should be able to tell you it has four lines of symmetry (the diagonals are lines of symmetry). The order of rotational symmetry is four because a square fits into its outline four times when it is rotated once.

5 Parallelogram

A parallelogram can be made in several ways.

- Draw on both sides of a ruler and then join the lines with two parallel lines.
- Put two congruent triangles together. One triangle is turned round.

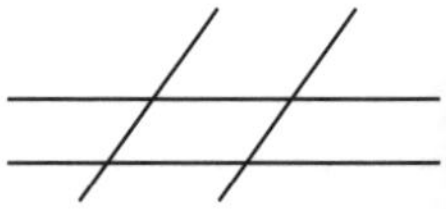

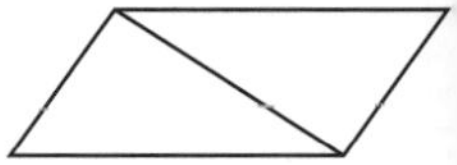

What can the pupils tell you about the sides of a parallelogram?

- Opposite sides are parallel and equal.
- Opposite angles are equal.
- There are no lines of symmetry.
- It has an order of rotational symmetry of two.

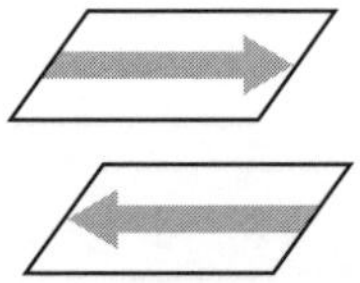

6 Rhombus

A rhombus can be made from two identical isosceles triangles put base to base. (An isosceles triangle can be made from two identical right-angled triangles.)
So four identical right-angled triangles are needed, made from two identical rectangles.

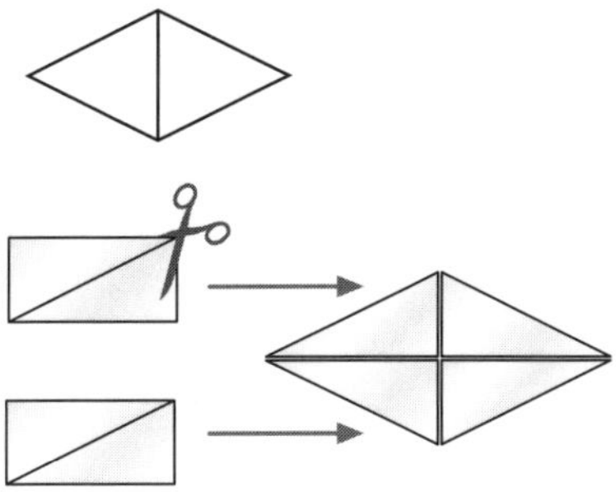

Stick the four triangles together and the pupils can find the properties of a rhombus.

- All four sides are equal.
- Opposite angles are equal.
- Diagonals bisect at right angles.
- The diagonals are the lines of symmetry.
- The order of rotational symmetry is two.

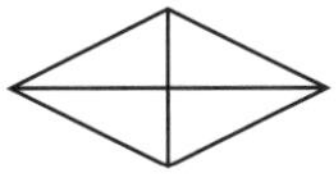

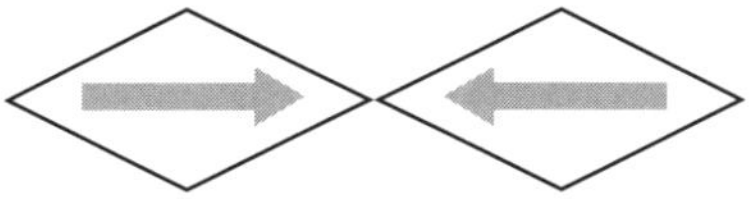

7 Kite

Pupils will be familiar with kites. They are a common toy. Like the rhombus, a kite can be made from two pairs of isosceles triangles. The triangles have the same base but should have different heights.

The pupils can use their models to find the properties of a kite.

- Two pairs of equal sides.
- Diagonals meet at 90°.
- There is one line of symmetry.
- The order of rotational symmetry is zero.

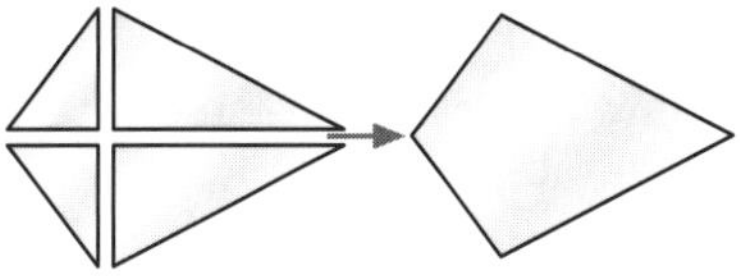

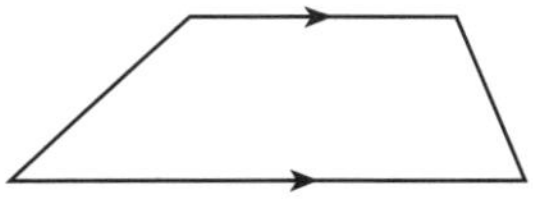

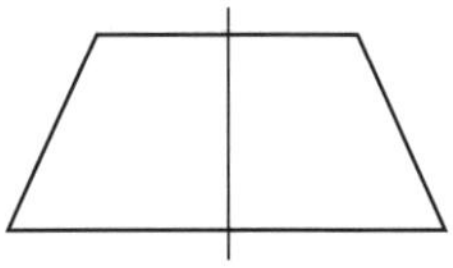

8 Trapezium

Two parallel lines, of different lengths, joined together make a trapezium. A special case is the **isosceles trapezium.** In this case the two sides that are not parallel are equal. An isosceles trapezium has one line of symmetry.

> ***Activity***
>
> - Try making the shapes using the methods suggested. Do you know any other ways of making the shapes?
> - Why are kites shaped the way that they are? Would a rectangle, a parallelogram or a rhombus not fly so well?
>
> Discuss the relationships between various shapes. Is a square a rectangle? Is a parallelogram a rhombus? Is a kite a trapezium? Give reasons for your answers.

Topic 42 Length and perimeter

1 Introduction

There are several stages to the development of children's understanding of length and its measurement:

- learning the vocabulary of length,
- comparing the lengths of objects,
- measuring the lengths of objects using non-standard units,
- estimating and measuring the lengths of objects using the metric system.

It is important that children do a lot of practical work whilst doing this topic.

> ***Activity***
>
> Think of as many words for length and distance as you can. Give examples of their use in sentences.
>
> What are the translations of these words into the local languages where you are?

2 Preliminary ideas and vocabulary

Before children can measure length, they will need to learn the language of length and distance. Words of length include 'long', 'short', 'tall', 'thick', 'thin', 'fat', 'breadth', 'broad', 'width' and 'wide'. Words of distance include 'near' and 'far'. Ask the pupils to use these words in sentences and describe objects as long or short, thick or thin, etc.

3 Comparing lengths

Let the pupils use the correct vocabulary to compare lengths. They can compare two objects: longer, shorter, taller, thicker, thinner, fatter.

They compare three or more objects: longest, shortest, tallest, thickest, thinnest, and fattest.

Pupils can compare many things just by looking at them. The classroom is obviously longer than a desk in it, and so on. Most pupils are shorter than their teacher. But what about objects that are roughly the same length? Children can compare them by putting them **next to each other**. Is my pencil longer than your pencil? Who is taller, you or me? How can we find out? You can ask pupils to stand in order, from tallest to shortest.

These activities teach the meanings of the words, and the ideas involved, in an interesting way.

Pupils need to understand **conservation** of length. The length of an object does not change when it is moved.

These lines are the same lengths:

So are these:

Some pupils will think that the bottom line is longer, even if they have seen it move from the first position. (Demonstrate using two equal length sticks or rulers.)

> ***Activity***
>
> *A short walking stick is longer than a long ruler.*
>
> Discuss this sentence.

4 Length using non-standard units

Children cannot always compare things by putting them next to each other. Some things cannot be moved, such as the length and breadth of a room. We need some way of measuring. Begin with arbitrary, informal methods.

- The length and breadth of a room can be measured by counting the number of **paces** needed to walk across it, the number of **jumps** or **arm spans** needed to cross it.
- The length of a desk can be measured by counting the number of **hand spans**, the number of **pencils** or pens needed to go across the desk.

There are other possible ways to measure: the cubit (the distance from the elbow to the tips of one's fingers), paperclips (for the length of a book). Discuss why we get different measurements if we use different length pencils, for example.

> ***Activity***
>
> - Devise activities for measuring lengths using non-standard units – paper clips, paces, cubits, etc. Try them out with your colleagues.
> - Discuss what to do with 'the leftover bit' when you measure. Suppose a desk is three and a bit hand spans long. How do you describe the bit left over?

5 The metric system

The children will find that they get different answers for the length of a room using strides. Larger pupils will get smaller answers. This will bring out the necessity for an agreed standard length that we can all use everywhere. The standard we use is the **metre**.

> **Note:**
> People everywhere now use the metric system developed over 200 years ago in France. It is simple and easy to use in calculations. The complete system is:
>
> 10 millimetres (mm) = 1 centimetre (cm)
> 10 centimetres = 1 decimetre (dm)
> 10 decimetres = 1 metre (m)
> 10 metres = 1 decametre (dam)
> 10 decametres = 1 hectometre (hm)
> 10 hectometres = 1 kilometre (km)
>
> This is a large set of units and, in practice, the important measures are the millimetre, centimetre, metre and kilometre. The relationship between these is quite simple:
>
> 10 millimetres = 1 centimetre
> 100 centimetres = 1 metre
> 1000 metres = 1 kilometre

The size of commonly used measures can be illustrated with these and other examples:

- A pin is about the same width as one millimetre, so 1000 pins side by side make about a metre.

- The smallest finger of an adult is about the same width as a centimetre. (Get your pupils to find out which of their fingers, or thumb, is about 1 cm wide.)
- A running stride is about the same length as a metre, so 1000 strides are about 1 km.

Many pupils have problems using these units and decimals. For example, 1 km 24m may be written as 1.24 km (when it should be 1.024 km) and 3.15 km is thought to be 3 km 15 m (when it should be 3 km 150 m). Draw attention to this.

6 Measuring and estimating

The best way to learn about measuring is to actually measure things. Pupils should learn to measure accurately. They should learn to estimate lengths and then compare their estimates with their measurements. This sort of experience will develop a true appreciation of the size of the various units.

Measuring with a ruler

Rulers are normally 30 cm long, and the centimetres are each divided into 10 millimetres. The centimetres are numbered, but the millimetres are not, although the 5th millimetre is drawn longer to help in counting the number of millimetres.

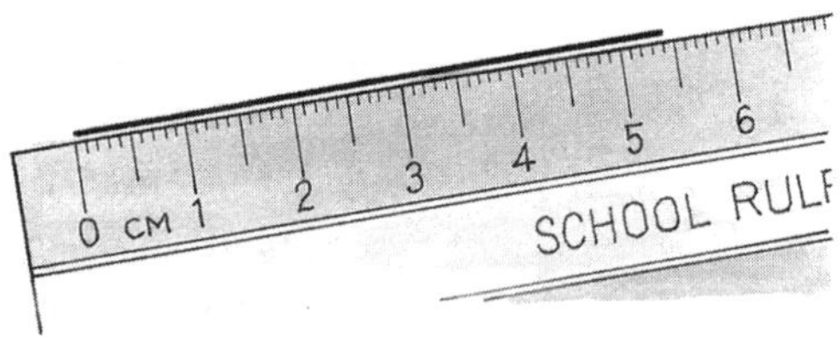

Pupils need to learn to start measuring from 0 and not from 1 or the end of the ruler. Some children get confused between mm (with large numbers) and cm (small numbers) on the ruler.

Start by getting the pupils to measure to the nearest centimetre. They will soon realise there is a question of accuracy — ask the pupils what to do with the part of a centimetre left over. They may suggest ignoring it or choosing the nearest centimetre or half centimetre. Later on, they can become more accurate and measure to the nearest millimetre. Then measurements can be reported in two ways. A line 4.8 cm long is also 48 mm long. When the children know about decimals they can use either.

Lines that are not straight

Which line is the shortest? Which is the longest?

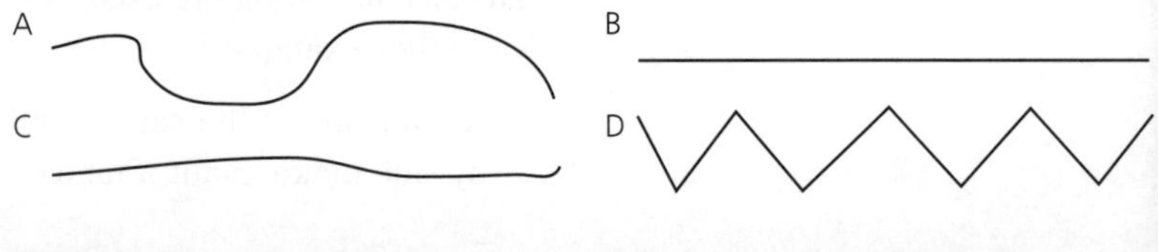

> ***Activity***
>
> Here are some suggestions for length-measuring activities. Try them out with your colleagues. Decide whether they are suitable for your pupils.
>
> - A standing jump for each pupil.
> - How far up a wall each pupil can reach (and perhaps compared with his/her height).
> - Distances in classrooms and lengths of furniture in them. Use tape measures and draw plans from the measurements.
> - The length of the stride of each pupil in the class, and find the mean.
> - The thickness of some paper (or other thin objects) by putting ten or more of them side by side, finding their total width and dividing by ten.
> - Distances on the sports ground and around the school using a tape measure or trundle wheel.

Measure the lines with a piece of string. Other curved things you can measure are the distance round pupils' knees, ankles, and wrists, distances on a map, etc.

Measuring long distances

Try to give pupils a chance to measure long distances. This requires a tape measure or a trundle wheel. The sports department may have these.

Trundle wheels have circumferences of exactly 1m. They are used for measuring distances such as the length of a football pitch. For each metre that a person pushing a trundle wheel walks, the trundle wheel goes round once and makes a 'click.' The number of clicks is the number of metres walked.

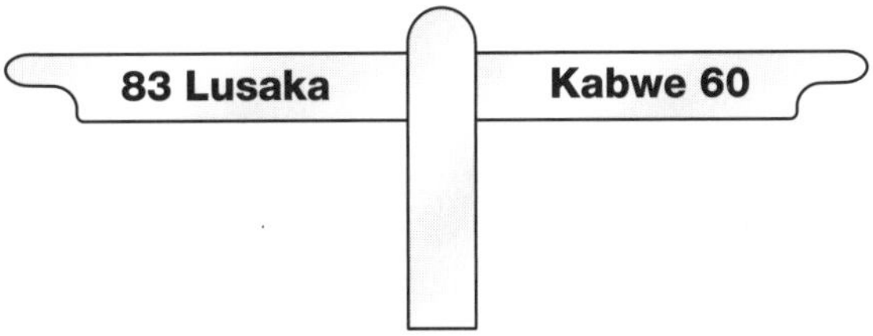

Pictures of **signposts** are a good source of problems concerning distances. For example, if a signpost says Lusaka is 83 km away and Kabwe is 60 km away in the opposite direction, how far apart are Lusaka and Kabwe?

7 Perimeter

Many pupils learn the formula $P = 2(l + b)$ for the perimeter of a rectangle, but do not understand where it comes from. This formula is one that does not help pupils. It uses brackets that many pupils find difficult. It also only applies to rectangles and not to other shapes.

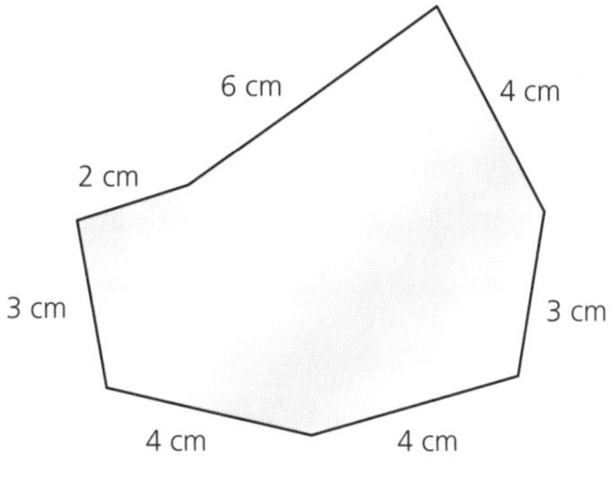

A better approach would be to consider the perimeter of all polygons. Start with a numerical example. To introduce a discussion of the perimeter of a shape, ask how far an ant will walk if it walks around the shape. The answer is to add up the lengths of all the sides of the figure. This distance around a shape is its perimeter. To find the perimeter of figures like this we do not need a formula; simply by adding the lengths of the sides we find the perimeter is $3 + 2 + 6 + 4 + 3 + 4 + 4 = 26$ cm.

The pupils should have lots of practice in finding the perimeters of shapes like this by adding the lengths of the sides together. This will reinforce in their minds the meaning of the word 'perimeter'.

By looking at a number of examples like this, it will be seen that the perimeter of a rectangle is length + breadth + length + breadth.
Here the perimeter = $4 + 6 + 4 + 6 = 20$.

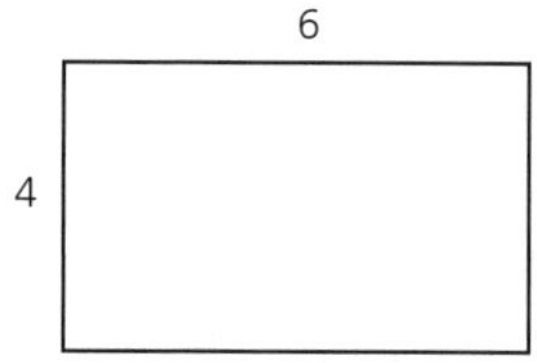

For many pupils, it is best to leave the result like this. To reach the formula, note that pupils can work this out by adding length and breadth and multiplying by 2.

$6 + 4 = 10$; $10 \times 2 = 20$

After a few examples like this some pupils will be able to see that the perimeter of a rectangle can be found by adding length and breadth together and multiplying by 2.

The perimeter of a rectangle with length l, width w is

$l + w + l + w = 2l + 2w = 2(l + w)$.

Topic 43 Area of a rectangle

1 Introduction

There are several stages to the development of children's understanding of area and its measurement:

- learning the vocabulary of area,
- comparing the areas of shapes using non-standard units,
- estimating and calculating the areas of shapes. The shapes include the rectangle, square, triangle, rhombus, kite and parallelogram.

2 Preliminary ideas and vocabulary

Area is a measure of the size of the surface (or space covered) of an object or shape. Before children can calculate areas, they will need to understand this and to learn the language of area. Words for area include 'big', 'small', 'larger'.

3 Comparing areas

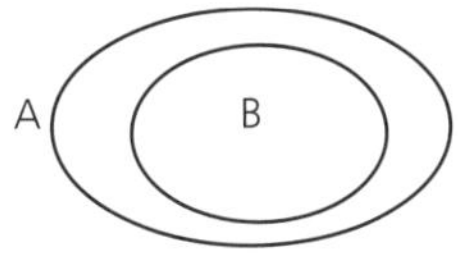

Start by comparing areas of different shapes by putting them on top of each other. If one shape completely covers the other it has a larger area:

Shape A has a larger area than shape B.

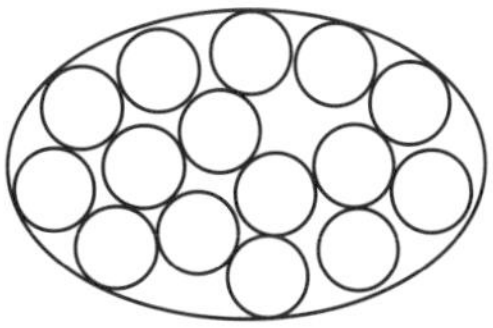

15

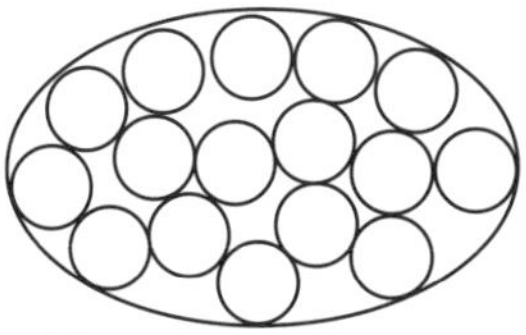

16

When the areas are similar it is hard to compare. Put small counters, coins, etc. on them and count the number of counters.

There is a problem with the gaps between counters. This would be reduced with smaller counters but they would be harder to count.

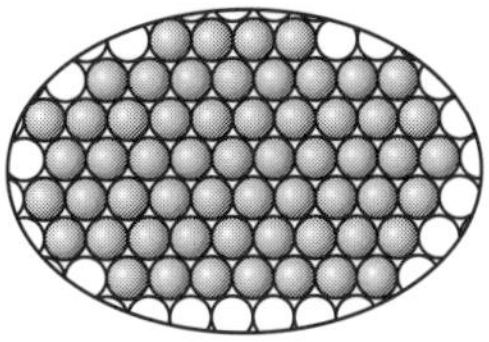

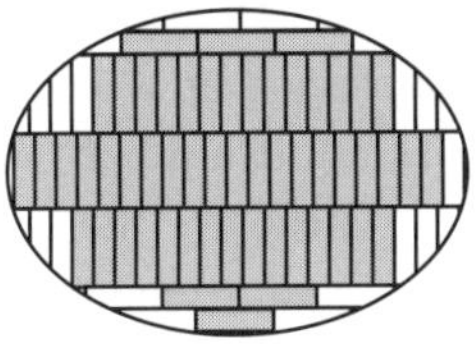

Other possible units you could use to measure area are old postage stamps, matchboxes and bottle tops. Gaps make the answer inaccurate. Two of these measures, postage stamps and matchboxes, do not leave gaps, and are less inaccurate, (but there are still problems if the shape whose area you are trying to find is rounded).

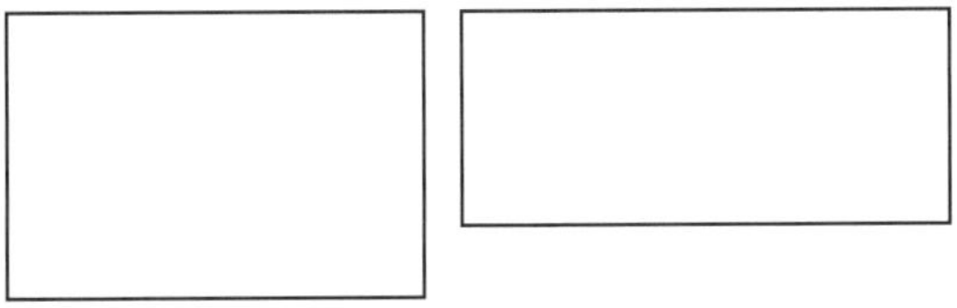

There are several shapes that do not leave gaps, but the shape most commonly used to measure area is the square. Squares are easy to count.

For example, the areas of these shapes can be compared by dividing them up into 1 cm squares. (The shapes are 6 cm by 4 cm and 7 cm by 3 cm and they have the same perimeters.)

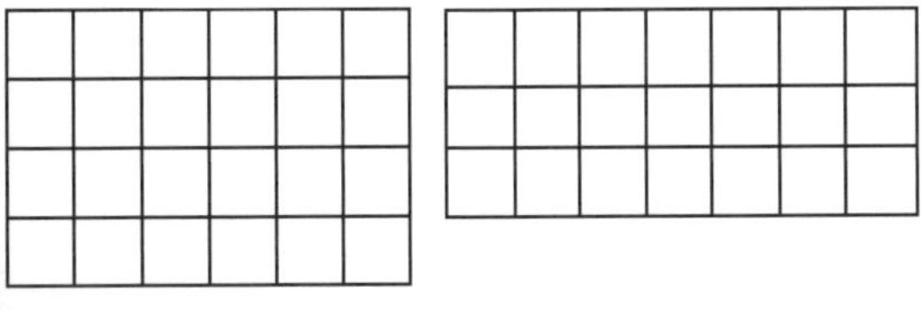

Counting the squares, there are 24 and 21 squares respectively, so the left hand shape is larger in area (even though it is shorter).

Activity

Collect together and use suitable small objects (bottle tops, matchboxes, etc.) to find and compare the areas of various objects or diagrams.

Some shapes cover objects with patterns (called tessellations) which leave no gaps. These shapes include rectangles, equilateral triangles, hexagons and parallelograms.

Consider the reasons why the square is easiest to use.

4 Formulae for the area of a square and a rectangle

Pupils will find it easiest to find a formula for area when the shape is a rectangle. Get them to draw rectangles on squared paper, and divide them up into one-centimetre squares. For example, this is a 6 cm by 3 cm rectangle.

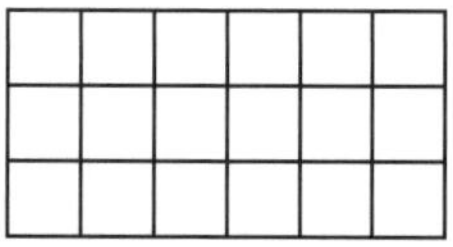

The rectangle has been divided up into 18 squares.

It has an area of 18 square centimetres, or 18 cm^2.

The pupils can find the areas of other rectangles and the teacher can put a table on the board like this:

Length	Width	Area
6	3	18
6	4	24
5	2	10

Ask the pupils what they notice.
The pupils should be able to say **without being told** that

area = length × width or **area = length × breadth**

Using letters, this becomes $area = l \times w$. Do not use letters at this stage. Most pupils do not like algebra very much and are more comfortable with words.

An important special case is when the shape is a square. A square is a rectangle with equal length and width, so

area of a square = length × length = (length)^2.

(In letters, if the length of a side is a cm, then the area is a^2 cm^2.)
Now the pupils can do examples using the formulae.

5 Area in the metric system

When the pupils have understood how to find the area of a square, they can understand the relationship between the commonly used units of the metric system. These are the square millimetre (mm^2), the square centimetre (cm^2), square metre (m^2) and hectare (ha).

Note: A hectare is the area of a square 100 m by 100 m. The basic unit of area in the metric system is the 'are' which is the area of a square 10 m by 10 m. (It is not used very much.)

$$1\,cm^2 = 1\,cm \times 1\,cm = 10\,mm \times 10\,mm = 100\,mm^2$$
$$1\,m^2 = 100\,cm \times 100\,cm = 10\,000\,cm^2$$
$$1\,ha = 100\,m \times 100\,m = 10\,000\,m^2$$
$$1\,km^2 = 1\,km \times 1\,km = 1000\,m \times 1000\,m = 1\,000\,000\,m^2 = 100\,ha.$$

Activity

Investigate traditional ways of measuring area that were used a hundred or more years ago in your area.

Topic 44 Areas of irregular shapes, triangles and quadrilaterals

1 Introduction

This topic considers how to find the areas of shapes other than rectangles and squares.

2 Irregular shapes

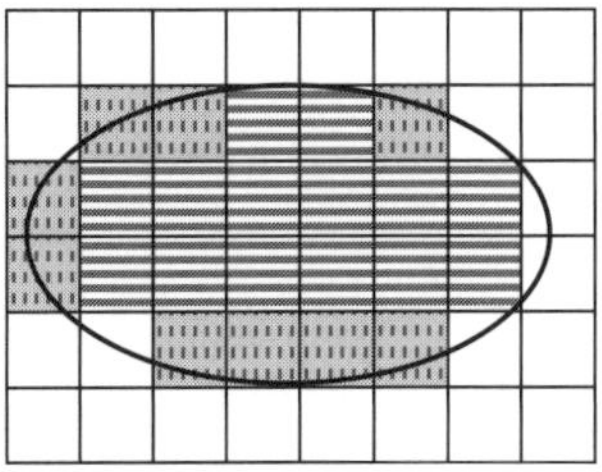

The area of this shape whose edge is curved can be **estimated**. Each square in the grid has an area of 1 cm^2. To find the area of the shape you have to count the squares covered by it. Squares that are completely covered by the shape are counted as 1 cm^2. Squares that are at least half-covered by the shape also count as 1 square centimetre. Squares less than half-covered are ignored.

In the shape there are 14 whole squares (shaded ▤). There are also 9 squares (shaded ▦) that are at least half inside the ring.

The white pieces of squares are ignored.
An estimation of the total area of the shape is 14 + 9 = 23 cm^2.
Pupils can practise this method by drawing shapes on squared paper and counting the squares that are shaded or at least half-shaded.

3 Area of a triangle

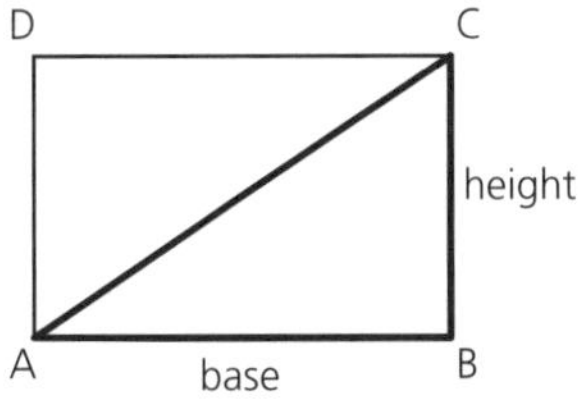

To find the formula for the area of a triangle, start with the simplest type, a right-angled triangle. A right-angled triangle is half a rectangle, and therefore has half the area of the rectangle.

Area of triangle ABC is half the area of rectangle ABCD.
ABCD has area AB × BC.
So, the area of triangle ABC = half of AB × BC.

The area of a right-angled triangle is $\frac{1}{2}$ × base × height.

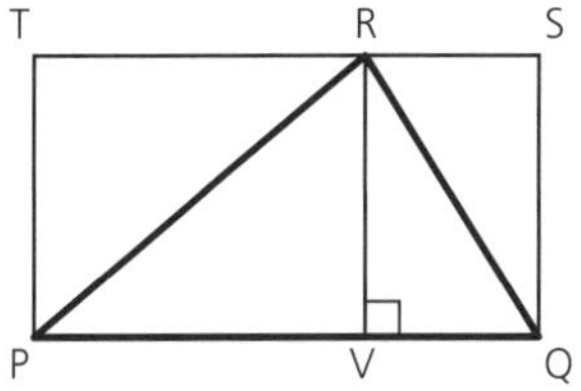

In fact, any triangle can be divided into two right-angled triangles. In the diagram, triangle PQR is in rectangle PQST. The height of the triangle, VR = the height of the rectangle, QS.

area of triangle PQR = area of PVR + area of QRV
area of PVR = $\frac{1}{2}$ × area rectangle PVRT
area of QRV = $\frac{1}{2}$ × rectangle VQSR
So, area of PQR = $\frac{1}{2}$ × area rectangle PVRT + $\frac{1}{2}$ area rectangle VQSR
= $\frac{1}{2}$ × area of the large rectangle PQST
= $\frac{1}{2}$ × PQ × RV = $\frac{1}{2}$ × **base** × **height**

Pupils will probably 'see' from inspection that the area of ΔPQR is half the area of PQST, and will not require this formal proof.

4 Area of a parallelogram

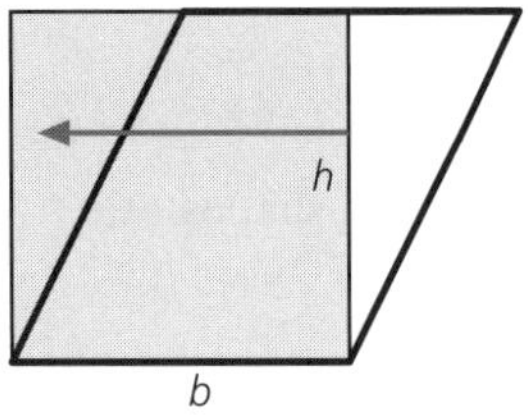

There are 2 ways pupils can find this.

- By moving a triangle from one end of the parallelogram to the other, the parallelogram is made into a rectangle.
  The area of the rectangle is base × height.
  The rectangle and parallelogram have the same area,
  so,
  area of a parallelogram = base × height

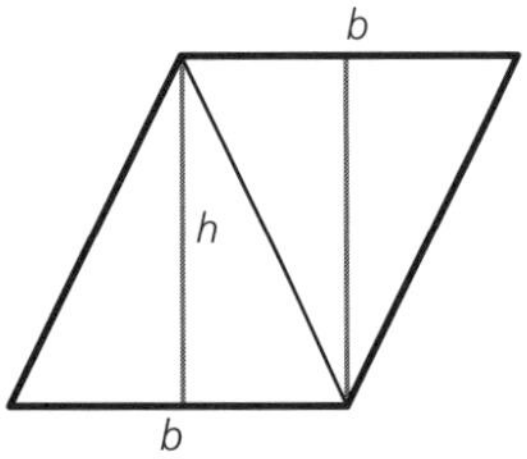

- A parallelogram can be turned into two identical triangles. Each triangle has an area of $\frac{1}{2}$ × base × height.
  So, area of parallelogram = 2 × $\frac{1}{2}$ × base × height = **base** × **height**

5 Area of a rhombus

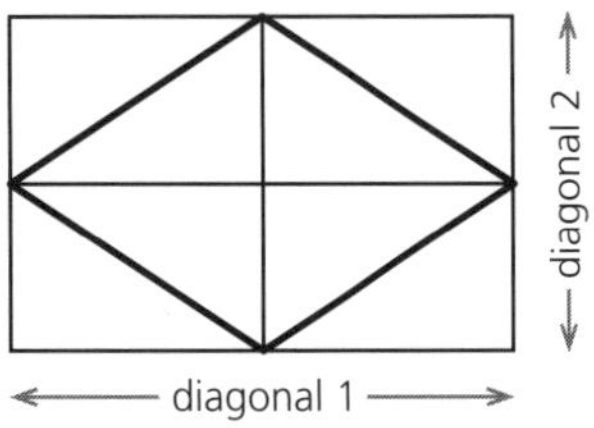

A rhombus is a parallelogram with equal sides so the formula for its area is the same as the formula for the area of a parallelogram.

area of a rhombus = base × height

The area of a rhombus can also be written in terms of the lengths of its diagonals.
The area of the rhombus is half the area of the containing rectangle.

area of rectangle = $l \times w$ = diagonal 1 × diagonal 2
area of rhombus = $\frac{1}{2}$ × area of rectangle
= $\frac{1}{2}$ × **diagonal 1** × **diagonal 2**

6 Area of a kite

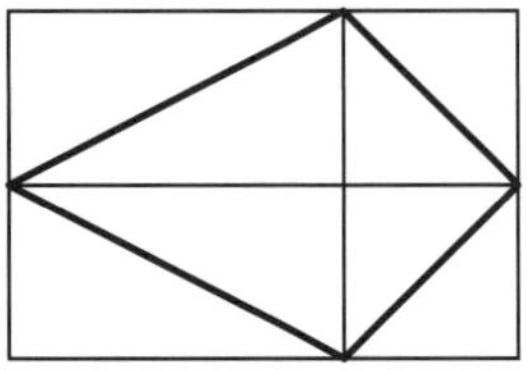

The same argument can be used to show that the area of a kite is also $\frac{1}{2}$ × **diagonal 1** × **diagonal 2.**

7 Area of a trapezium

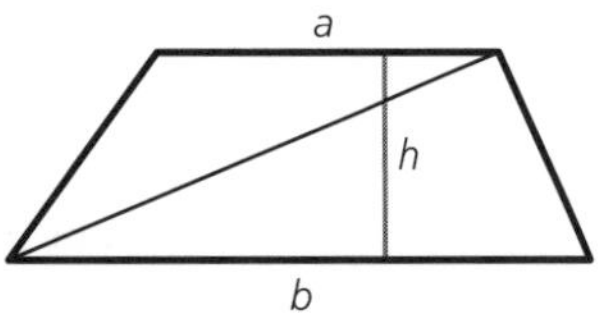

A trapezium has a pair of parallel opposite sides.
The sides are length a and b. They are h apart.
The trapezium can be divided into two triangles.

area of top triangle = $\frac{1}{2}$ × base × height = $\frac{1}{2}\, a\, h$
area of bottom triangle = $\frac{1}{2}$ × base × height = $\frac{1}{2}\, b\, h$

So, area of trapezium = sum of areas of the two triangles
= $\frac{1}{2}\, a\, h + \frac{1}{2}\, b\, h = \frac{1}{2}\, \mathbf{(a + b)\, h}$

8 Composite shapes

After they have learnt the formulae for specific shapes, show the pupils a composite shape like the one at the top of page 201. How can they find its area? They should suggest dividing it into two rectangles. Their areas can be worked out, and then added. There are also some missing lengths to work out. The shape can be divided up in different ways, into A and B, or C and D.

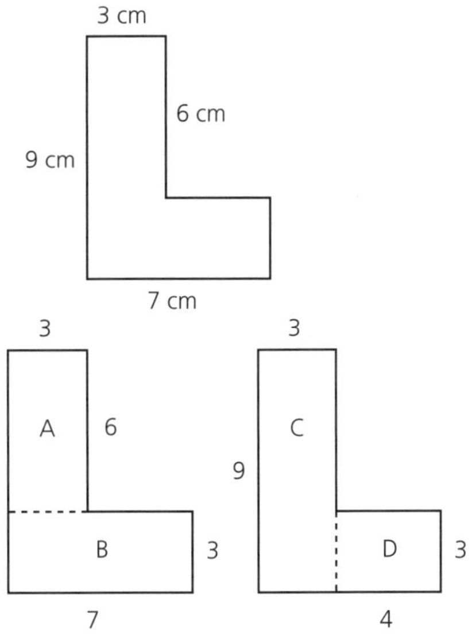

area A = length × width = 6 cm × 3 cm = 18 cm^2
area B = 7 cm × 3 cm = 21 cm^2

So, total area = 18 cm^2 + 21 cm^2 = 39 cm^2

Or: area C + area D = 9 cm × 3 cm + 4 cm × 3 cm
= 27 cm^2 + 12 cm^2 = 39 cm^2

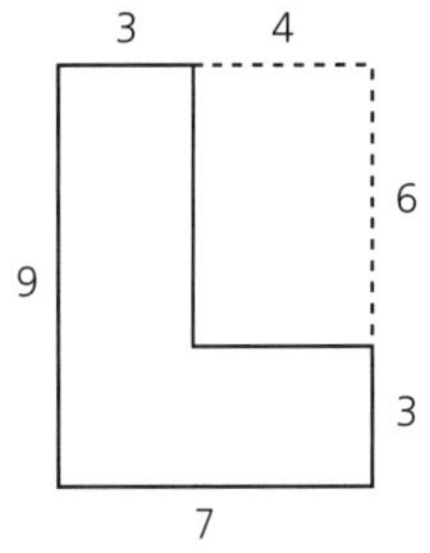

Working out an answer in two ways, and getting the same result shows the pupils that there is more than one method of finding the area.

There is a third way that this can be done: by subtracting areas.

area of shape
= area large rectangle – area of small rectangle
= 7 cm × 9 cm – 4 cm × 6 cm = 63 cm^2 – 24 cm^2 = 39 cm^2

Pupils need to choose the most effective way of calculating the area in each case. The example of an L shape illustrates in a simple way the choices possible in these calculations.

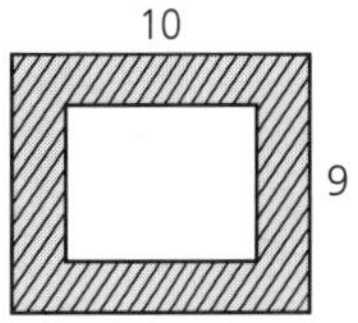

Area word problems often involve finding the area of a path around a rectangular lawn.
These are usually most simply solved by finding the difference in areas.
Pupils often forget that the difference between the outer lengths and the inner lengths is *twice* the width of the path. In the drawing, if the path is 1 m wide, the lawn is 10 – 2 = 8 m by 9 – 2 = 7 m, and not 9 m by 8 m.

Activity

Find out about Pick's theorem which states that if a polygon is drawn on dotted paper, with vertices on the dots, then the area of the polygon is given by the formula

$$A = \tfrac{1}{2}O + I - 1$$

where O is the number of dots on the outline of the shape and I is the number of dots inside the shape.

In the example, $O = 8$, $I = 4$, so $A = 7$.

Consider teaching this as a problem solving exercise.

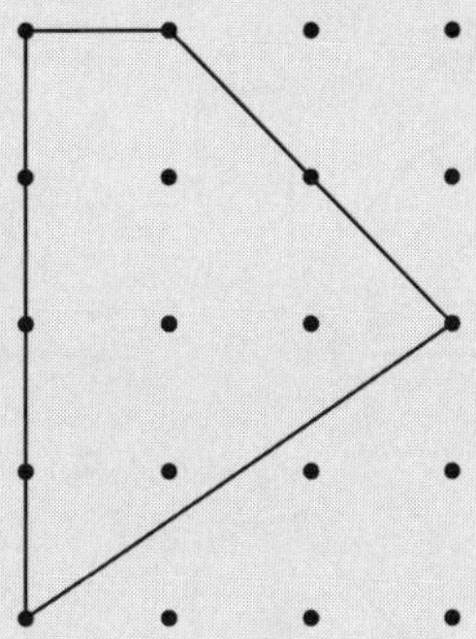

Activity

Try some problem solving exercises like this:

- You are wrapping parcels. You have a book 16 cm × 25 cm × 3 cm.
  How much overlap do you need at the edges and around the middle?
  How large a sheet of paper should you use?
  How much paper would you need for a parcel of 5 books?
- What size of paper would you need to make an envelope 10 cm by 15 cm?
  Would different designs of envelope (but the same size) need different amounts of paper?

Topic 45 Solids

1 Introduction

Pupils have to learn about many solid objects. Books can be used as examples of cuboids, empty tins as examples of cylinders. Oranges are spheres. Ant-hills are often conical.

2 Nets

It helps pupils, and makes learning more interesting, if they make model cubes, and so on, for themselves. This topic looks at the nets of a few solids. The net of a solid is made from all the faces of the solid put together on a flat piece of paper so that when folded it makes the required solid.

Cube

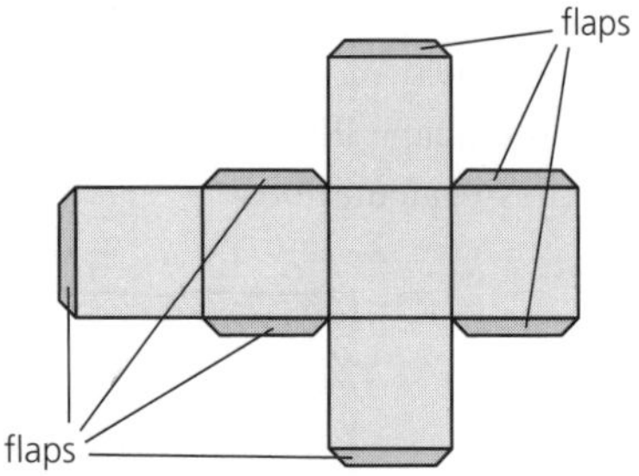

A cube has 6 identical square faces. They have to be joined together. One arrangement is four in a row and one on each side The easiest way to draw this shape is on squared paper. The squares can be made any size you like. If this shape is cut out you will need to join the edges with sticky tape. If you add flaps to the net, they can be used to glue the shape together.

Cuboid

The cuboid is a solid whose faces are all rectangles and the sides are at right angles to each other. It will have a net similar to the cube. Any grocery box will give the pupils an example of the net of a cuboid.

Other shapes

Pupils will enjoy making other solids, such as **pyramids** and **prisms**. Discuss with your pupils the names of the faces from which they are made. A pyramid has a flat base and sides that are isosceles triangles. The sides of a prism are rectangles and the ends can be any flat shape. The nets (without flaps) look like this.

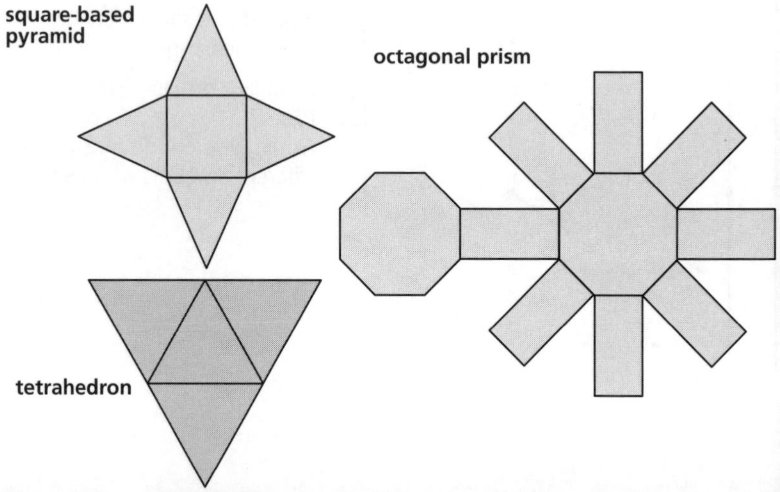

3 Euler's law

This law is an interesting result about the faces, edges and vertices of a solid. It can be discovered by pupils who are asked to count the number of faces, edges and vertices of different solids. They need to know that the **faces** are what make up the net, the **edges** are the lines joining the faces and the **vertices** are the points where three or more faces meet. The pupils can build up a table by looking at various shapes with straight edges:

Solid	No. of faces	No. of vertices	No. of edges
Cube	6	8	12
Cuboid	6	8	12
Tetrahedron	4	4	6
Square based pyramid	5	5	8
Octagonal prism	10	16	24

With the correct questioning, pupils should be able to discover that the pattern is

number of faces + number of vertices = 2 + number of edges

This is known as Euler's law.

Topic 46 Volume and capacity

1 Introduction

Piaget distinguished between the 'interior volume' of a box and the volume that a box occupies. This is similar to the difference between capacity and volume. The capacity of a container is the amount that the container can hold. The volume of a solid is the amount of space that it fills. The main metric units for volume and capacity are litres (l), millilitres (ml), cubic metres (m^3) and cubic centimetres (cm^3). This topic looks first at the teaching of volume and then at capacity.

There are several stages to the development of children's understanding of volume and its measurement:

- learning the vocabulary of volume,
- comparing the volumes of shapes using non-standard units,
- estimating and calculating the volumes of objects such as cubes, cuboids and prisms.

Work on volume follows on from work on length and area.

2 Preliminary ideas and vocabulary

Volume is a measure of the size of a solid. The same words are used to compare volumes (big, small, larger, etc.) as were previously used for areas. These words need to be used in the context of volumes.

3 Comparing volumes

Give the pupils some plasticine or clay to work with. Ask them to make it into a ball, a box and other shapes. Ask them if shape is the same. (No.) Is the amount of clay the same? (Yes.) The amount of 'stuff' used to make the object is the same. The 'volume' (not to be confused with the volume of a radio) has not changed.

Next, compare a large object and a small object by putting them side by side. It is easy to state which is larger. However, when objects are about the same size, or different shapes, it is harder to compare them. In some cases, we can compare how much they can hold. To compare two boxes or containers fill the boxes with sand, small pebbles, beads, beans, or rice. (These are less messy than water and can be used in objects that need to stay dry.) Compare the amount in the two boxes. This can be done in several ways.

- Fill one box and then to pour the filler into the other box and see if it overflows. Pupils can estimate how many times they can fill a large container from a small one, and then check their estimate using sand or another filler.
- Count the number of pebbles or beans needed to fill each box.

The aim is for pupils to be able to compare the volumes of things accurately. Stones may be of different sizes. The number of pebbles and so on can be counted but there is a problem with the gaps between pebbles. This is reduced if you use something like flour or sand.

4 Finding the volume of cubes and cuboids

Large rectangular-prism-shaped boxes could be filled with bricks. They have no gaps, and the number of bricks can be counted. Small boxes can be filled with little cubes. Pupils will find one-centimetre cubes hard to make. However, they can make 2 cm or 5 cm cubes and use them. The pupils can count the cubes.

The pupils are now ready to explore the number of cubes needed in a more precise way. They will find it easiest to find a formula for volume when the shape is a cuboid. They do not need to put the cubes in a box; the cubes can stand up on their own.

If you do not have enough small cubes, other things will do: matchboxes, empty tin cans, etc.

Give groups of pupils 40 or 50 cubes. Ask them to use some of the cubes to make a rectangle.

Ask the pupils to make some more rectangles that are identical, and put them on top of the first rectangle.

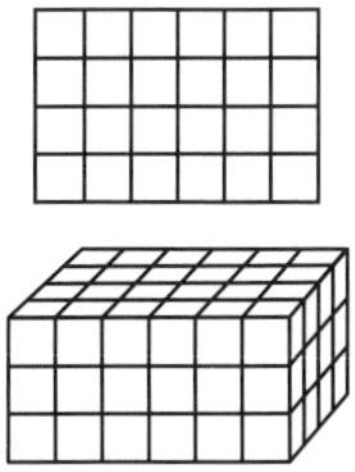

In the diagram there are three layers of 6×4 cubes. How many cubes are used? The pupils can count them and find that the 3 by 6 by 4 cuboid has 72 cubes. They can also realise that each layer has 24 cubes (6×4). There are 3 layers so the number of cubes is $3 \times 24 = 72$.

(Note that length, width and height were all different in the example. This was deliberate, to avoid confusion.)

The area of a shape is the number of 1 cm squares needed to cover it.

The **volume** of a solid is the number of 1 cm cubes needed to fill it

A few examples of this type should show the pupils that the

volume of a cuboid = length × width × height.

Pupils can deduce that a cube is a special case of a cuboid:

volume of a cube = length × length × length = (length)^3.

5 Volume in the metric system

When the pupils have understood how to find the volume of a cube, they can understand the relationship between the cubic millimetre (mm^3), the cubic centimetre (cm^3) and the cubic metre (m^3).

$$1\,cm^3 = 1\,cm \times 1\,cm \times 1\,cm = 10\,mm \times 10\,mm \times 10\,mm = 1000\,mm^3$$
$$1\,m^3 = 100\,cm \times 100\,cm \times 100\,cm = 1\,000\,000\,cm^3$$
$$1\,km^3 = 1\,km \times 1\,km \times 1\,km = 1000\,m \times 1000\,m \times 1000\,m = 10^9\,m^3$$

6 Volume of a prism

The idea of a prism can be introduced using slices of bread that are all the same shape. The slices cut across the loaf are the **cross-section** of the loaf. All the cross-sections are the same. A prism is a solid that has the same cross-section throughout its length.

A cuboid is a prism whose cross-section is a rectangle. When the pupils know that the volume of a cuboid = length × width × height, they can use this knowledge to find the formula for the volume of any prism.

area of cross-section of cuboid = width × height
volume of cuboid = area of cross-section × length

We can develop the formula for other prisms. A triangular prism can be thought of as half a cuboid.

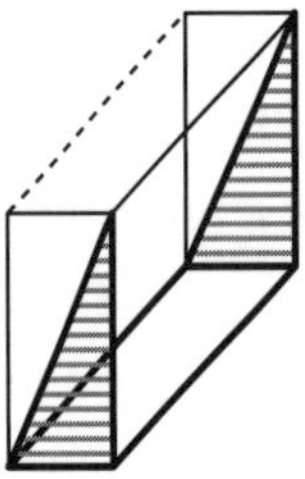

volume of the cuboid = length × width × height
volume of the triangular prism = $\frac{1}{2}$ × length × width × height

Note also that:

area of cross-section of triangular prism = $\frac{1}{2}$ × width × height
So, volume of the prism = area of triangle × length
= $\frac{1}{2}$ × width × height × length

It is true that for any prism:

volume = cross-sectional area × length
or **= area of base × perpendicular height**

Volume of a cylinder

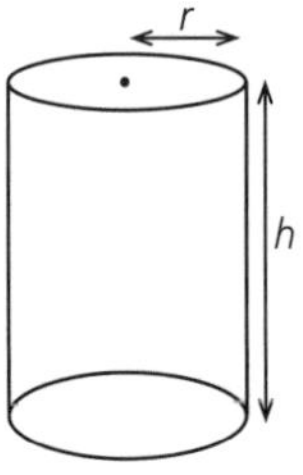

A cylinder is a commonly used prism. It has a circular cross-section. If the circular cross-section has a radius r and the height of the cylinder is h, pupils should be able to state that:

area of the circular base = πr^2
and **volume of a cylinder** = $\pi r^2 \times h$.

See Section E, Topic 47 for a more detailed consideration of circles.

7 Capacity

The capacity of a container is a measure of the amount it can contain. The metric units of capacity are the millilitre and the litre.

1000 millilitres (m*l*) = 1 litre (*l*)

The units of volume and capacity are closely connected.

1 m*l* = 1 cm^3 1 litre = 1000 cm^3

Doctors use the 'teaspoon' as a measure for giving liquid medicines. A teaspoon can hold about 5 ml (5 cm^3) of liquid. Liquids that are sold in bottles are often sold in litre bottles. (Try to get some to show to the class.)

> ***Class activity***
>
> This requires three bottles marked at one litre. Mark a scale on each bottle in 100 m*l* sections.
>
> Use two bottles to measure two amounts of liquid. Add them together in the third bottle.
>
> Estimate whether a container holds one litre. Then test to see if the estimate is correct.
>
> This activity should develop a sense of how much a litre is.

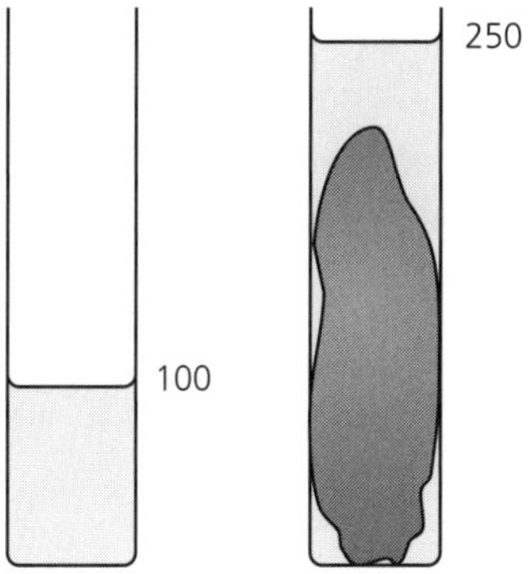

8 Irregular objects

The picture illustrates how to find the volume of an irregular object. The measuring cylinder has 100 m*l* of water in it. A stone is put in the water and the water level rises to the 250 m*l* mark. This means that the volume of the stone must be 150 m*l*.

The pupils can use this method to find the volumes of any objects that are not damaged by being put in water. They should do this with objects whose volumes cannot easily be calculated using a formula, as well as to confirm the volumes of objects whose volume can be calculated.

Topic 47 Circles

1 Introduction

Pupils will be very familiar with circular objects, such as balls, wheels, tins and pipes. Huts are often circular; they enclose the largest area within a particular wall length. The teacher can use this familiarity to begin discussion of circles.

2 Language of circles

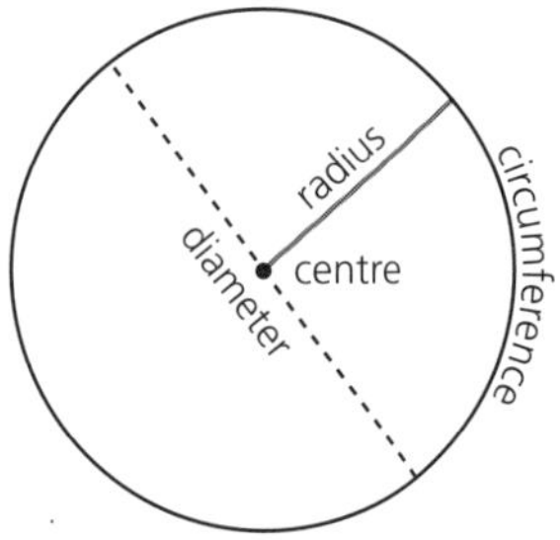

Pupils need to learn the vocabulary of circles. The main words are **centre**, **radius**, **diameter** and **circumference**. All these words are used both for the lines and for the lengths of the lines. A simple way to show a circle is to draw around an empty cylindrical tin.

Another way is to use a piece of string held at its ends by two pupils. One pupil stands still, the other moves round making sure that the string is tight. This can also be done with a pencil on a piece of paper. The length of the string is the radius.

3 Relationship between radius, diameter and circumference

The relationship that the diameter is twice the radius is simple, and easy to establish. The diameter is the distance across a circle, which is two radii.

Finding the relationship between the circumference of a circle and its diameter is more difficult. Ask the pupils which is longer, the circumference or the diameter? They will agree that the circumference is longer, but how much longer is it?

An estimate to the answer to this question requires a collection of different size old tins. Pupils can be asked to bring these to school. The pupils should be split into small groups and each group given a tin, a ruler and a piece of string. If string is unavailable strips of paper or strands from a creeper can be used. The diameter can easily be measured using the ruler. Ask the pupils how they can find the circumference. They should be able to say that they can put the string or paper around the tin. The pupils can then measure the circumference and diameter of their tins. The teacher can then put a table of the pupils' results on the board. It will look something like this:

Circumference	24.9 cm	39.0 cm	21.3 cm	51.5 cm	16.5 cm
Diameter	8 cm	12.4 cm	7 cm	16 cm	5.2 cm

Ask the pupils to find the difference between each circumference and diameter. A discussion should quickly lead to a realisation that there is not a constant difference between diameter and circumference. Ask the pupils to think about how the two measurements might be related. It should seem natural to investigate what $\frac{\text{circumference}}{\text{diameter}}$ is.

This will lead to the table looking like this:

Circumference	24.9 cm	39.0 cm	21.3 cm	51.5 cm	16.5 cm
Diameter	8 cm	12.4 cm	7 cm	16 cm	5.2 cm
$\frac{C}{D}$	3.11	3.15	3.04	3.22	3.17

The pupils will notice that all the answers are similar. The teacher can now tell the pupils that circumference ÷ diameter is always a little more than 3. In fact it is 3.14 to 2 decimal places. The number is called π (pronounced 'pi').

The teacher should emphasise that 3.14 is only an approximation. The real value for π is a decimal that goes on forever, 3.1415927… The other common approximation used in calculations is $\frac{22}{7} = 3\frac{1}{7}$. This approximation should only be used when the radius or diameter is a multiple of 7. Otherwise, use the decimal approximation.

Activity

The history of the attempts to find the value of π by the Greeks, Babylonians, Egyptians and Chinese is very interesting. Investigate this, or, better still, ask pupils to do it.

The relationship circumference ÷ diameter = π is usually written as:

circumference = π × diameter

or $C = \pi d$.

Since $d = 2r$, it is also written as $C = 2\pi r$. This form is useful when the radius is known and a problem asks pupils to find the circumference.

4 Area of a circle

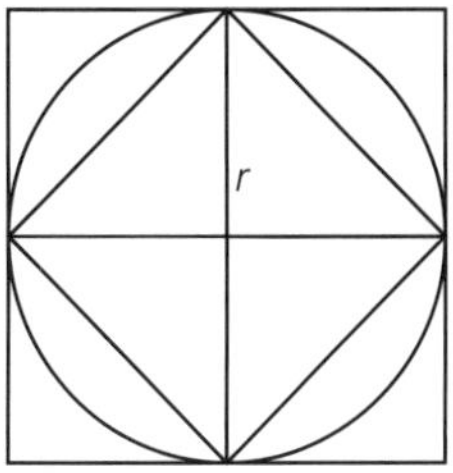

The teacher can show the pupils that the area of a circle is a value between $2r^2$ and $4r^2$ by drawing squares inside and outside the circle.

The area of the big square is $(2r)^2 = 4r^2$.
The big square is made up of 8 identical triangles. The small square is made up of 4 of these triangles, so the area of the small square is half of $4r^2$ or $2r^2$.

The circle is bigger than the small square and smaller than the big square, so its area is somewhere between $2r^2$ and $4r^2$.
The exact formula can be deduced from the formula for the circumference by drawing a circle and cutting it up into 16 sectors.

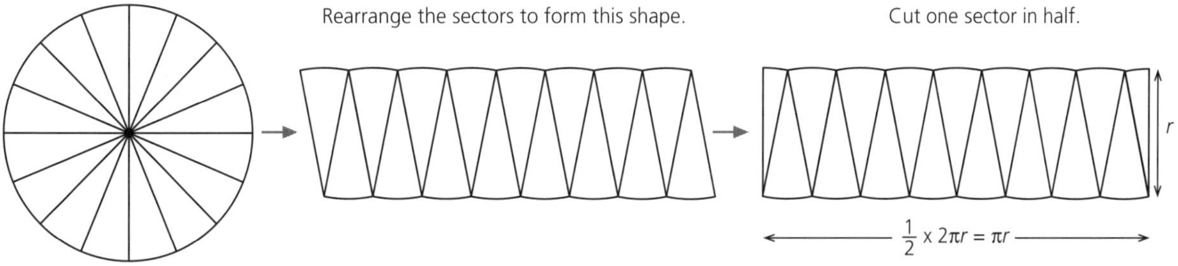

The pupils will agree that the shape is roughly rectangular. It can be made more rectangular by cutting off half of one end sector and moving it to the other end.

The circumference of the circle makes up the two lengths of the rectangle so the length of the rectangle is approximately half the circumference

$= \frac{1}{2} \times 2\pi r = \pi r.$

The width of the rectangle is approximately the radius of the circle, r.

Area of circle $\approx$ area of rectangle $\approx r \times \pi r = \pi r^2$

> ### *Activity*
> Prepare a teaching aid that demonstrates the area of a circle by this method.

5 Cylinder

The cylinder is a prism with a circular cross-section, so its **volume** is:

area of base × height = $\pi r^2 h$

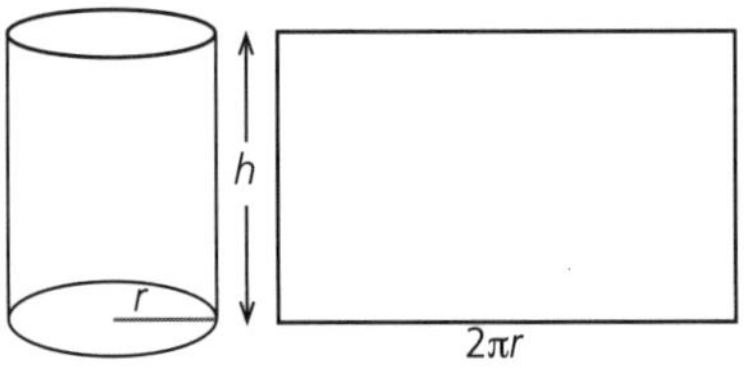

The **curved surface area** of a cylinder is most simply found by bringing a cylindrical tin to the class that still has its paper label on. This can be removed carefully and seen to be a rectangle with a length of $2\pi r$ and height h.

curved surface area = $2\pi r \times h = 2\pi rh$

The surface areas of the base and the top are both πr^2. So,

total surface area = $2\pi rh + 2\pi r^2 = 2\pi r\ (h + r)$.

6 Sectors

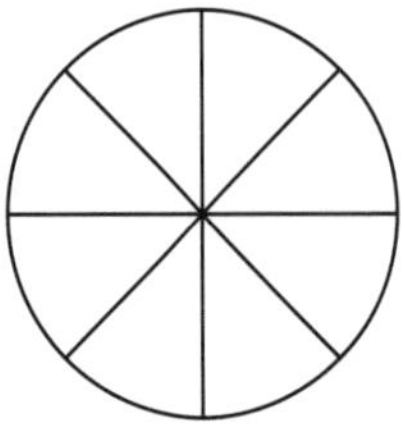

Cakes are usually cut up by cutting pieces from the centre.

These pieces are sectors.

Pupils can make sectors with the angle discs described in Section E, Topic 37.

Pie charts

Sectors are important in pie charts. (Point out that this 'pie' has nothing to do with 'π'.) Pupils have to realise that the area of a sector is proportional to the angle at the centre. As with so much else in maths, this is done by considering simple examples.

a quarter of a cake (one quarter of the area)

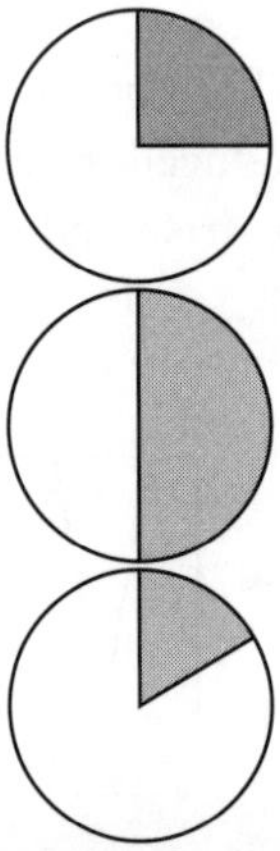

The angle of the sector, 90°, is a quarter of 360°.

a semi circle (one half of the area)

The angle of the sector, 180°, is half of 360°.

a sector angle 60° (one sixth of the area)

The angle of the sector is $\frac{1}{6}$ of 360°.

Examples such as these will lead pupils to realise that the area of a sector is the same fraction of the area of the whole circle, as the angle is of the whole circle (360°). That is: angle is (*fraction*) of 360°, so area of sector is (*fraction*) of area of circle.

(Primary children are too young for the intimidating formulae.)

Topic 48 Coordinates

1 Introduction

Coordinates are an important topic. They link algebra and geometry and can be introduced in an enjoyable way.

Coordinates in the classroom

Start with positive coordinates. Use the classroom seating arrangements for an effective introduction. Arrange the desks in the classroom neatly in rows and columns.

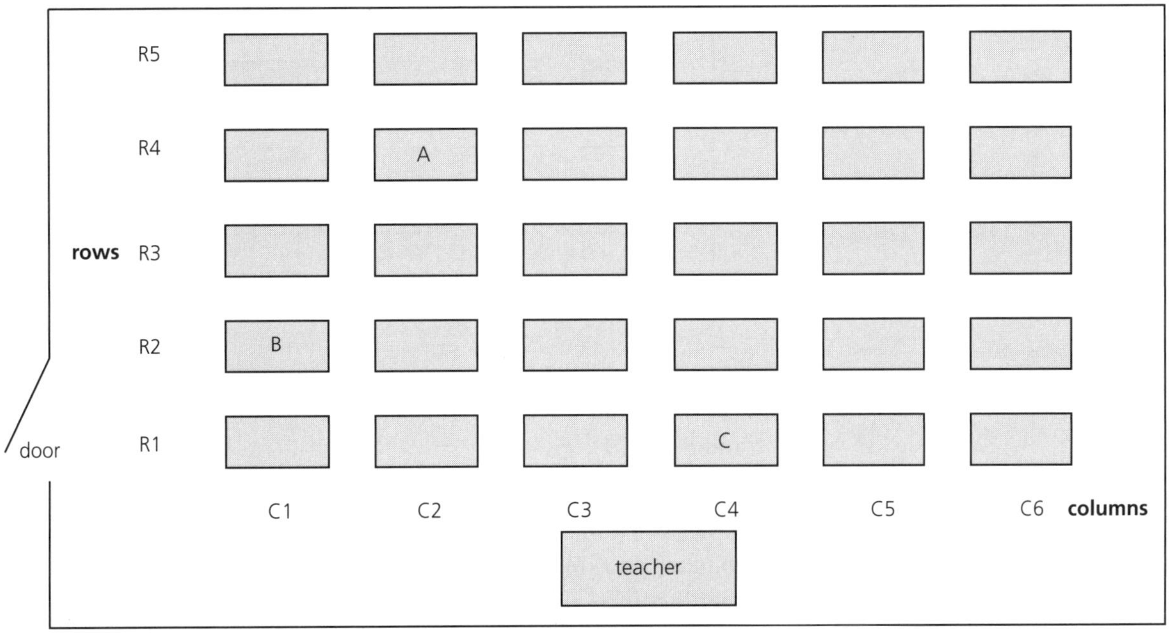

Give the pupils this problem:

> Suppose that you want your friend to get a book from your desk. How would you tell your friend which is your desk?

They should be able to say that they would describe the position of the desk by stating its row and column. Ask them to describe the position of their own desks. Desk A, for example is in the 2nd column and 4th row. (Pupils need to know the difference between **rows** and **columns**.)

This describes the position precisely. The door is the starting point. This is natural from the point of view of someone who is coming to the class to look for a desk. It also fits well with the way the **origin, O,** for coordinates is in the bottom left-hand position for positive numbers.

'2nd column and 4th row' is written more briefly as (C2, R4). Even more briefly it can be written as (2, 4). These two numbers are the **coordinates** of desk A. Everyone has agreed to write the column number before the row number to avoid confusion. Ask the pupils to tell you the coordinates of desks B and C. (B = (1, 2) and C = (4, 1). Ask all the pupils for the coordinates of their own desks.

Coordinates game
Ask the pupils to play 'I am ..., who is ...?'. The first pupil looks at the position of his/her desk and says, for example, '*I am (2, 3), who is (4, 5)?*' The appropriate pupil in position (4, 5) replies and the game continues around the class. [With thanks to Wong Pak Set.]

An alternative approach. Label the diagram like this:

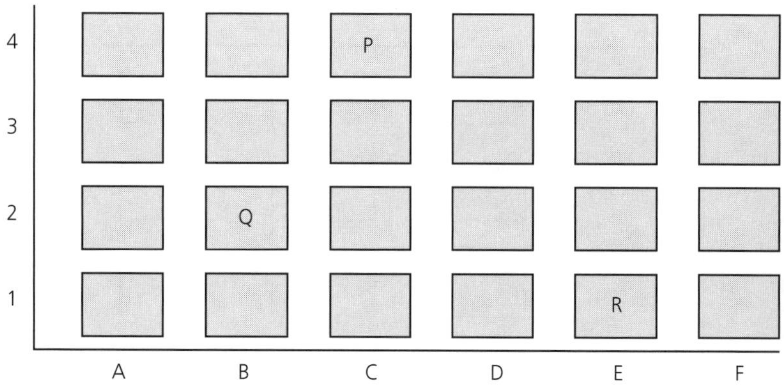

When writing the coordinates put the letters first so P = (C, 4), Q = (B, 2) and R = (E, 1). This leads easily into P = (3, 4), Q = (2, 2) and R = (5, 1).

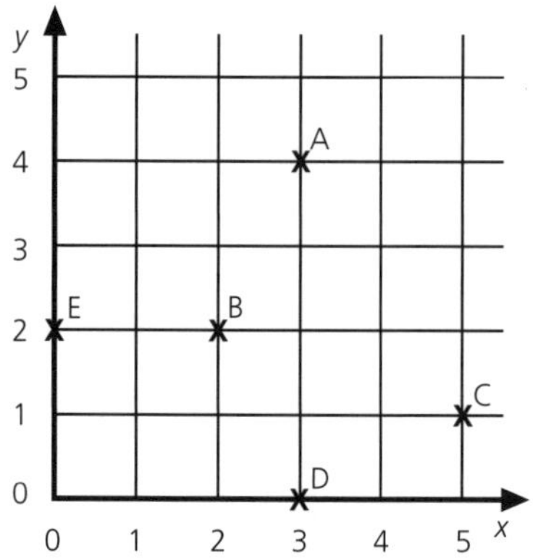

After the pupils have all learnt to identify the positions of their desks, you can draw the coordinates without desks and put lines for columns and rows. You can now introduce the language of coordinates — origin, axes.

The coordinates now include zeros.

The perpendicular lines along the bottom and side are called axes (singular:axis). They meet at the **origin, 0.** The vertical axis is the ***y*-axis**. The horizontal axis is the ***x*-axis.**
A = (3, 4), B = (2, 2), and C = (5, 1) as before.
D = (3, 0) and E = (0, 2) have zeros in them.

2 Equations of lines

Most topics in maths are more effectively taught in small classes. Unusually, for the approach suggested here, the larger the class the more effective it is. It can be done in the classroom or even with the whole school in assembly. The pupils are sitting in rows and columns. Make sure that they all know their own coordinates. Ask all the pupils whose x-coordinate is 1 to stand up. They will make a straight line from the front. This line is called $x = 1$.

Ask all the pupils whose y-coordinate is 2 to stand up. They will make a straight line across the class. This line is called $y = 2$.

Other examples can be given to illustrate the equations of other vertical and horizontal lines.

Then ask all the pupils whose y-coordinate is 1 more than their x-coordinate to stand up. This is another straight line across the class, with equation $y = x + 1$.

Do other examples, such as $y = 2x$. If the number of pupils is large enough you can show the graph of $y = x^2$. Pupils with coordinates (0, 0), (1, 1), (2, 4), (3, 9) and (4, 16) form a curve.

3 Negative coordinates

Once the pupils have learnt about negative numbers they can learn about negative coordinates. This is not a difficult extension. The axes are now in the middle of the diagram and not at the edge. Look at the coordinates of some points.

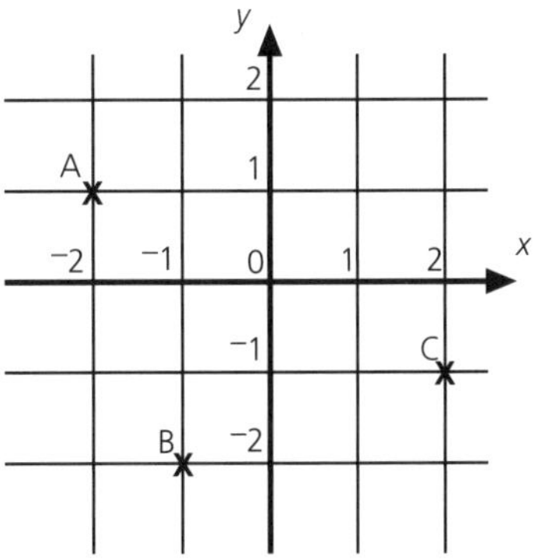

Pupils will be able to see that points in the top left of the graph, such as A, have a negative x-coordinate and a positive y-coordinate.

So A = ($^-2$, 1)

Points in the bottom left, such as B, have negative x- and y-coordinates.

So B = ($^-1$, $^-2$).

Points in the bottom right, such as C, have positive x- and negative y-coordinates.

So C = (2, $^-1$)

Activity

- Make a **plan** of your classroom using squared paper and coordinates. Here is a simple example.
- **Maps** Use coordinates to describe the positions of places. Find a map of your area. Prepare a lesson using coordinates to find the positions of towns and rivers with which your pupils are familiar.

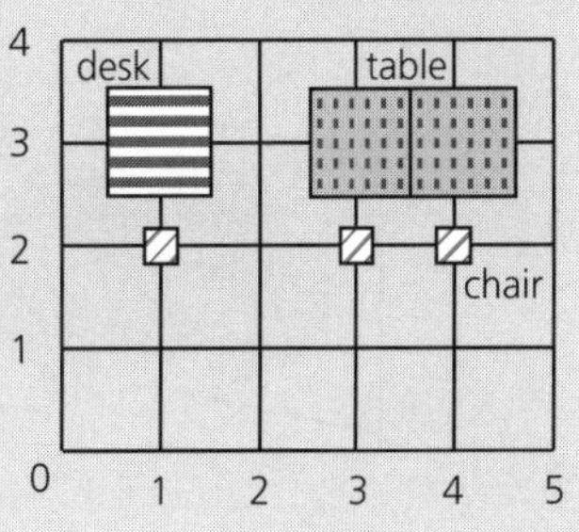

Topic 49 Line and rotational symmetry and tessellations

1 Introduction

Symmetry is very important in nature. Most creatures are symmetrical. Things that people make are often symmetrical. Cars are symmetrical to make them easier to drive. Kites are symmetrical.

There are two kinds of symmetry, associated with two types of transformation: reflection and rotation. If half an object is reflected about a line and it shows the original object, then it has **line symmetry**. If an object can be rotated and it still looks as if it is in its original position, then it has **rotational symmetry**.

2 Line symmetry

In the Primary school, the work on line symmetry should be informal and stress how an object or pattern can be seen as (or cut into) two identical but 'opposite' parts. Sometimes there is more than one way to make the 'cut', for instance with a square.

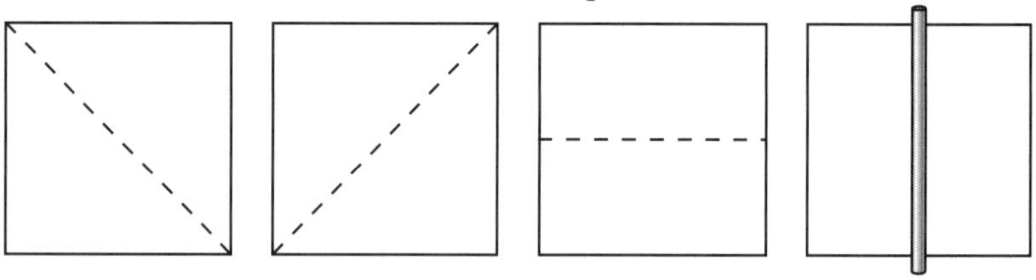

Ask the pupils to place a straw or straight stick along the line of symmetry.

Pupils can make symmetrical shapes in several ways:

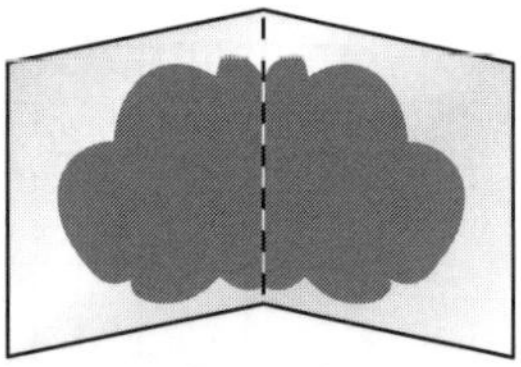

- **Ink blobs** Fold a piece of paper down the middle. Open it, put a small blob of ink or paint or mud in the middle. Fold the paper, smooth it down. Open the paper. There will be a pattern, like this one, that is symmetrical along the fold line.
- **Paper folding and cutting** This activity requires paper and scissors. Fold a piece of paper, cut out small pieces and then unfold the paper.

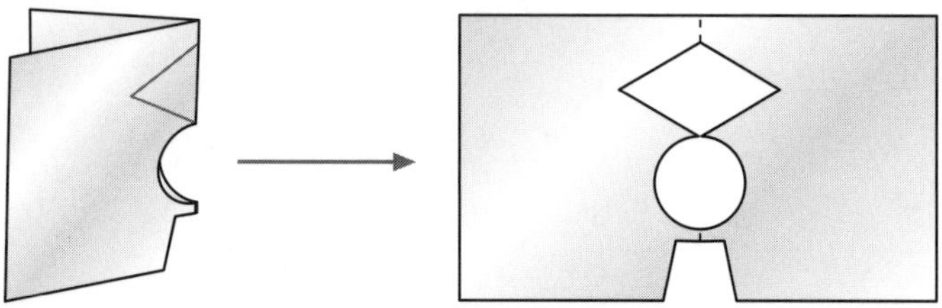

The fold in the paper is a line of symmetry. More intricate patterns can be made by folding the paper several times. Each fold will be a line of symmetry.

- **Carbon copies** Fold a piece of paper in four. Unfold it. Put a piece of carbon paper in the paper and fold it again. Draw a picture on one face of the folded paper making sure that the drawing goes up to line of the fold. Unfold the paper. The design is symmetrical.

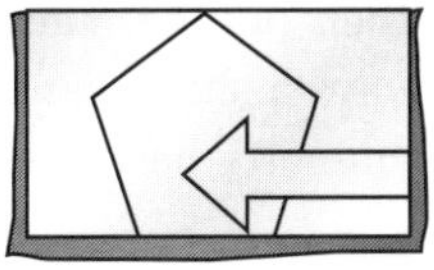

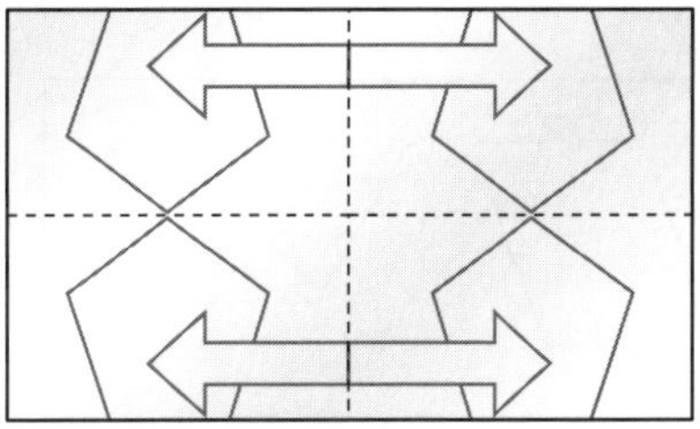

- **Using mirrors** Put a mirror on various pictures, patterns or letters to make symmetric shapes. Mirrors can also be used to test whether shapes are symmetrical by putting the mirror on the line thought to be the line of symmetry. If what is seen in the mirror is the same as when the mirror was not there then the shape is symmetrical about that line.

A mirror can also be used to disprove the widely held belief that the diagonals of a rectangle are lines of symmetry.

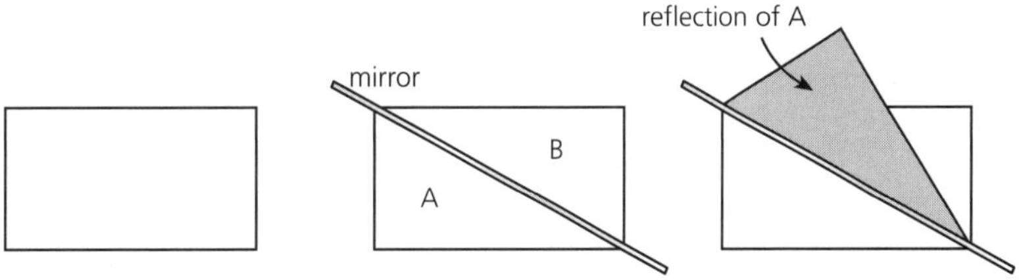

It can also be disproved by folding a rectangle along a diagonal.

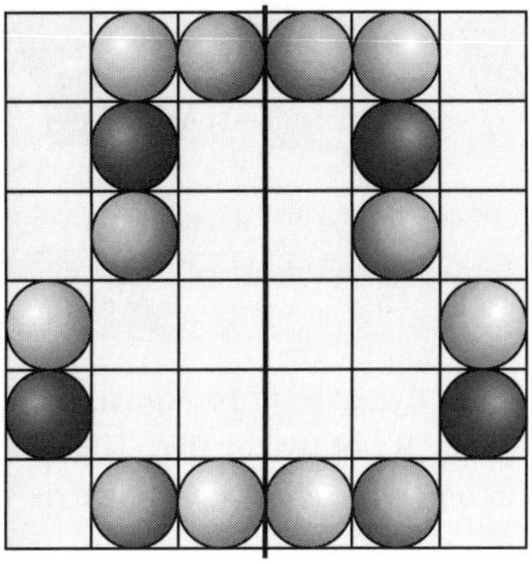

- **Egg trays** Put different coloured balls, or small objects, into an egg tray to make symmetrical shapes. Pegboards or thumbtack boards can be used in the same way.
- **Kites** Kites have to be symmetrical, or else they will not fly. The children can make a kite. The simplest way is to use the fact that the diagonals of a kite are perpendicular and they bisect each other (cut each other in half). The lightest material to use is old plastic bags.

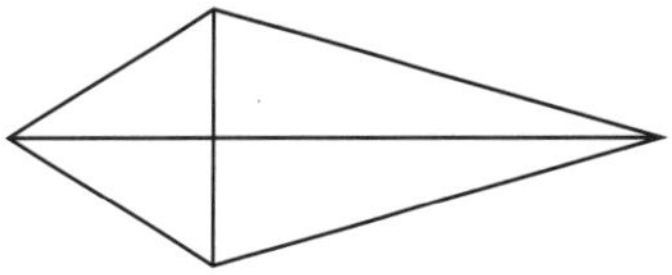

Activity

Line symmetry exercises: Prepare a collection of different half faces and other shapes that the children have to complete to make them symmetrical.

For example:

3 Rotational symmetry

Many things that children are familiar with have a turning pattern. When wheels, propellers, balls, some flowers, and so on, are turned around they still look the same.

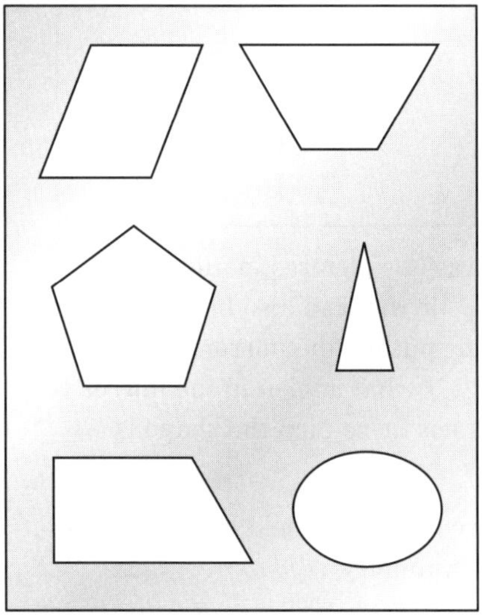

The shapes below can all be turned a part of a complete turn and still fit into their outline. The **order of rotational symmetry** is the number of ways a shape can fit into its own outline. A shape with no rotational symmetry has an order of rotational symmetry of 1.

Two of these shapes have rotational symmetry but no line symmetry, the others have both rotational and line symmetry.

Hole filling

Draw a shape on card and cut it out, leaving a hole of that shape in the card (see left). How many ways can it fit in the space it left? The shape will fit when it is turned around (rotational symmetry).

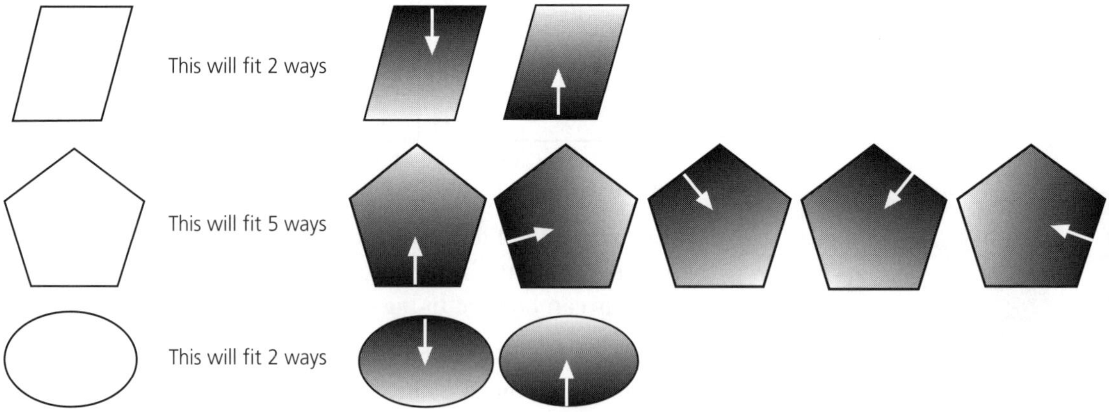

The shapes above are said to have two or five axes of rotational symmetry, or an order of rotational symmetry of two or five. The other shapes will only fit in the hole in one way. They have one axis of rotational symmetry.

You can also consider if the shapes would fit if they were turned over.

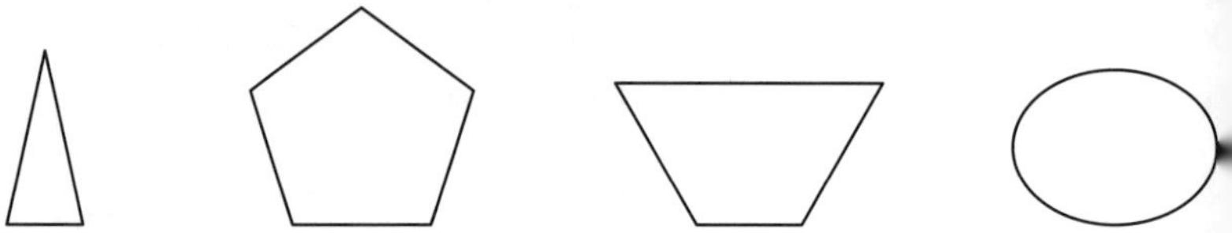

All these shapes would fit if they were turned over, and therefore have at least one axis of line symmetry.

Activity

Investigate the number of lines of symmetry, and the order of rotational symmetry of various shapes. Explore this work with your pupils.

Use cut-outs of shapes to arrange in symmetric patterns.

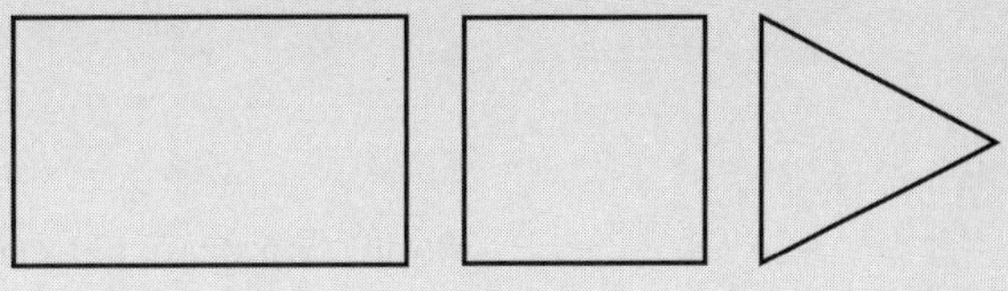

Investigate the symmetries of solid objects.

4 Tessellations

A shape tessellates if we can put many copies of it together in such as way that a plane surface is covered, with no gaps and no overlaps. For example, triangles, quadrilaterals and hexagons tessellate:

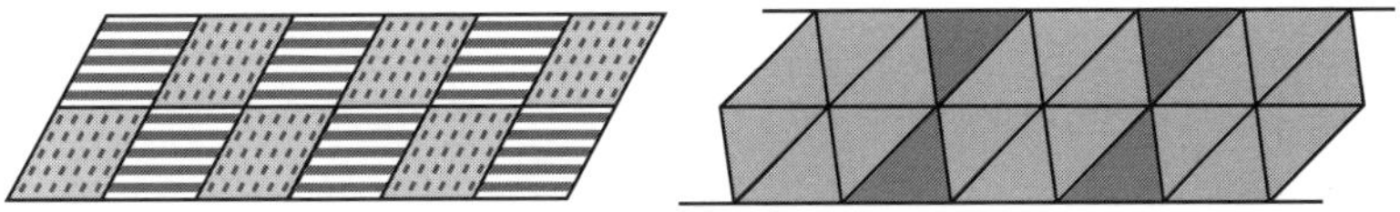

Pupils can use these basic designs to make very attractive patterns. They can colour different triangles different colours, for example. Their designs can be used to make the classroom more attractive.

A basic tessellation can be used to make more attractive patterns. For example, this is a simple tessellation of rectangles:

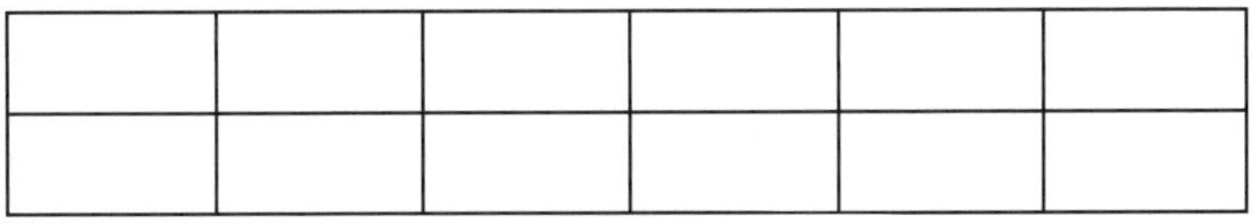

Cut one part off each rectangle and move to the other end:

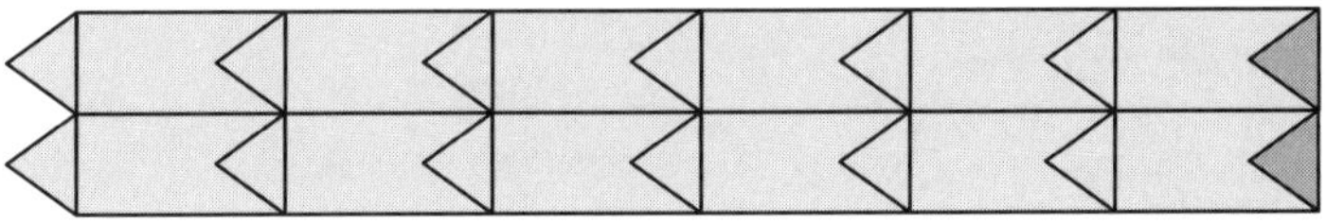

Or cut a curve out of each rectangle:

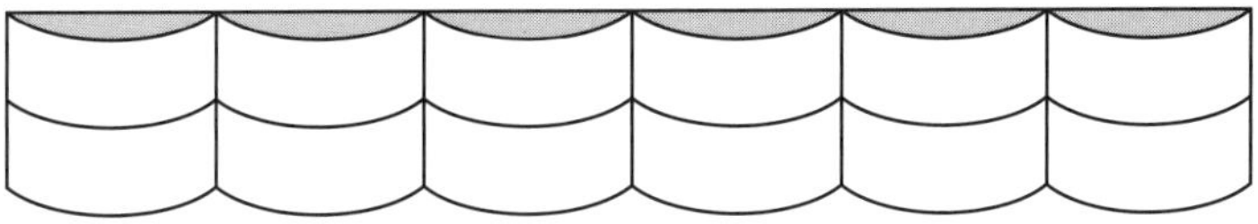

Or rotate part of the shape:

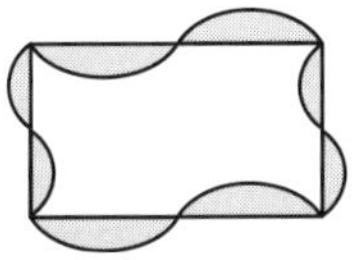

This gives rise to an attractive pattern.

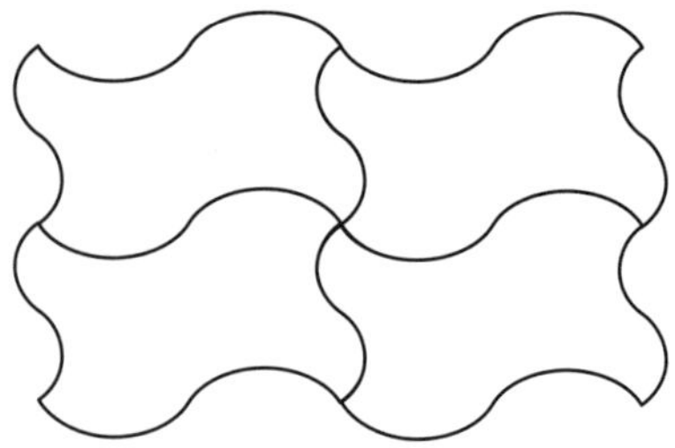

Pupils can draw patterns like these and colour them in or shade them. This is something they will enjoy. At the same time they will be learning how to draw and measure accurately, and improve their appreciation of the shapes used in geometry — triangles, rectangles, and so on.